PRINCIPLES AND PRACTICE

Springer

Berlin
Heidelberg
New York
Barcelona
Budapest
Hong Kong
London
Milan
Paris
Santa Clara
Singapore
Tokyo

M.R. Wilkins · K.L. Williams · R.D. Appel · D.F. Hochstrasser (Eds.)

Proteome Research: New Frontiers in Functional Genomics

With 36 Figures, 16 in Color

Springer

Dr. Marc R. Wilkins
Hôpitaux Universitaires de Genève
et Université de Genève
Laboratoire Central de Chimie Clinique
24, rue Micheli-du-Crest
1211 Genève 14
Switzerland

Dr. Ron D. Appel
Hôpitaux Universitaires de Genève
Division d'Informatique Médicale
24, rue Micheli-du-Crest
1211 Genève 14
Switzerland

Prof. Keith L. Williams
Australian Proteome Analysis Facility
and MUCAB
Macquarie University
School of Biological Sciences
Sydney
NSW 2109 Australia

Prof. Denis F. Hochstrasser
Hôpitaux Universitaires de Genève
et Université de Genève
Laboratoire Central de Chimie Clinique
24, rue Micheli-du Crest
1211 Genève 14
Switzerland

QP
551
, P756
1997

ISBN 3-540-62775-8 Springer-Verlag Berlin Heidelberg New York

CIP Data applied for
Die Deutsche Bibliothek - CIP-Einheitsaufnahme
Proteome research : new frontiers in functional genomics / ed. Marc R. Wilkins ... - Berlin ;
Heidelberg ; New York ; Barcelona ; Hong Kong ; London ; Milan ; Paris ; Santa Clara ; Singapore ;
Tokyo : Springer 1997
(Principles and practice)
ISBN 3-540-62775-8

© Springer-Verlag Berlin Heidelberg
1997 Printed in Germany

Camera ready by the editors
Cover design: D&P, Heidelberg using the illustration HepG2 2-D gel reference map from the
SWISS-2DPAGE database.
SPIN 10571786   39/3137 5 4 3 2 1 0 - Printed on acid free paper

To Sobet, Brynnie, Catherine and Anne-Catherine
To our children
To our parents

To Alan Williams and Matthieu Funk who would
have enjoyed this had they stayed around to experience it

Acknowledgments

This book is the fruit of a fantastic collaboration and friendship established around the world and in particular between scientists from Australia, Italy and Switzerland. We met in Sydney, Siena and Geneva, exchanged researchers and in 1996 at the Siena Proteome meeting conceived this book on Proteomics. There followed a division of duties that has put a heavy load on the researchers in our respective laboratories. All of us are busy people and we acknowledge the great contributions which our authors have made against tight deadlines!

Proteomic developments in Sydney have been possible thanks to the financial support of Macquarie University, the New South Wales State Government and the Federal Australian Government who together helped establish the first national proteome facility, the Australian Proteome Analysis Facility (APAF) in 1996. We also acknowledge the Australia Research Council and Australian Medical Research Council for supporting our research which has borne fruit as the proteome. The Sydney group acknowledges the unstinting support by many companies involved in scientific instrument developments and diagnostics. Without the support of Beckman Instruments (Australia and U.S.A.), Bio-Rad Laboratories (Australia and U.S.A.), GBC Scientific Equipment, Gradipore Limited, Hewlett Packard, Amrad Pharmacia Biotech, there would be no prototype instruments in our laboratory.

Proteomic developments in Geneva for the last fifteen years have been possible thanks to the financial support of the University of Geneva and University Hospital of Geneva, the Inter Maritime, Michelham, Montus and Helmut Horten Foundations, the Swiss National Fund for Scientific Research and several world-wide companies such as Electrophoretics plc, Bio-Rad Laboratories and Oxford GlycoSciences. The Geneva group also thanks Professors Alex F. Muller, Jean-Raoul Scherrer, Christian Pellegrini, Francis Waldvogel, Alain Junod and Robin Offord who have trusted our approach and supported our work for many years.

The first two proteome meetings in Siena in 1994 and 1996, which stimulated the realisation of this book, were possible thanks to Electrophoretics plc and the University of Siena.

Finally, we would like to thank Professor Edmond Fischer for his most supportive preface, which is the best encouragement we can get to further pursue our proteome work.

Marc Wilkins, Keith Williams, Ron Appel and Denis Hochstrasser

Preface

This is a timely book. It is being published as all the information stored in our DNA is about to be revealed, but demonstrates convincingly why this accumulated nucleic acid-based knowledge can only take us so far. The human genome might enable us to predict the proteins that can potentially be generated, but not where and when or at what level. It cannot tell us the cells in which proteins will be expressed or at which stages of development or differentiation this will happen. Nor can it take into account the enormous diversification of structure that results from alternate splicing, gene insertion, switching by deletions and recombination and other kinds of rearrangements as seen, for instance, in the immune system. Gene structure alone tells us very little about the physiological function of proteins since it ignores the co- and post-translational modifications to which they are subjected, such as their processing by limited proteolysis, their glycosylation, prenylation, ADP-ribosylations etc., and, of course, reversible phosphorylation.

The regulation of cellular processes requires a myriad of commands, positive and negative, to keep under control all reactions that take place and to make sure that no crucial event occurs at an inappropriate time or place. And we know today that the predominant signal that orchestrates these reactions, that turns the switches on and off, relies on protein phosphorylation. This is present to such an extent that at least 30% of all the proteins found in a mammalian cell extract exist in a phosphorylated state, even though one wouldn't know whether all these phosphorylation reactions are physiologically significant.

But even if we had all this knowledge at hand, one wouldn't know how cells react to given signals or, in multicellular organisms, synchronise their behaviour in response to internal or external demands. The modified proteins that are ultimately generated are only the words used by cells to transduce their signals or communicate with one another. For sentences to be constructed or for cross-talks to be initiated, proteins must interact with target, adapter or docking proteins and cells must interact with cells. This requires a plethora of binding motifs (e.g. the src homology SH2 and SH3 domains, or the PH, WW, PDZ domains), some no longer than a few amino acid residues, and further special sequences that determine the localisation of proteins to subcellular elements, direct their translocation in or out of the nucleus, their insertion into the endoplasmic reticulum or the plasma membrane or target them for destruction after internalisation. Cell-cell interaction brings into play an intricate, combinatorial system of cell adhesion molecules, crucial to the establishment of such highly sophisticated networks of communication one finds, for instance, in the immune system or the far more complex central nervous system where more than a billion cells speak with one another through a million billion synapses, ultimately leading to the generation of thought, memory and consciousness.

In all, then, no matter how distinct proteins might be from one another, they must be viewed as mosaics of structural elements, sorts of necklaces made up of a variety of beads in which each bead carries its own characteristic properties. However, it is the sum of these beads, their choice and particular disposition within the peptide chain that will ultimately determine the overall architecture and function of a protein.

This book, written and edited by those who conceived the notion of proteomics and contributed the most to its development, is the first to offer a comprehensive perspective of the field. It describes authoritatively its origin, fundamentals and background; its methodologies and how the data collected should be analysed, stored, retrieved and applied by the research and industrial scientist as well as the clinician. It eloquently documents how far these new technologies have taken us and where they might lead us in the future.

June 19, 1997

Edmond H. Fischer
Prof. Emeritus of Biochemistry
University of Washington
1992 Nobel Laureate in Medicine

Contents

List of Contributors

Ron D. Appel
Molecular Imaging and Bioinformatics Laboratory, Geneva University Hospital - D.I.M, 24 rue Micheli-du-Crest, CH-1211 Geneva 14 Switzerland. E-mail: ron.appel@dim.hcuge.ch

Amos Bairoch
Department of Medical Biochemistry, University of Geneva, 1 rue Michel Servet, CH-1211 Geneva 4 Switzerland. E-mail: bairoch@medecine.unige.ch

Luca Bini
Department of Molecular Biology, University of Siena, Pian dei Mantellini 44, I-53100 Siena Italy. E-mail: pallini@unisi.it

Andrew A. Gooley
Australian Proteome Analysis Facility (APAF) and Macquarie University Centre for Analytical Biotechnology (MUCAB), School of Biological Sciences, Macquarie University, Sydney NSW 2109 Australia. E-mail: agooley@rna.bio.mq.edu.au

Nicolas Guex
Geneva Biomedical Research Institute, Glaxo Wellcome Research and Development, 14 chemin des Aulx, Case Postale 674, CH-1228 Plan-les-Ouates Geneva Switzerland. E-mail: ng45767@ggr.co.uk

Ben R. Herbert
Australian Proteome Analysis Facility (APAF) and Macquarie University Centre for Analytical Biotechnology (MUCAB), School of Biological Sciences, Macquarie University, Sydney NSW 2109 Australia; Wool Research Organisation of New Zealand (WRONZ), Private Bag 4749, Christchurch New Zealand. E-mail: bherbert@rna.bio.mq.edu.au

Denis F. Hochstrasser
Central Clinical Chemistry Laboratory and Department of Medical Biochemistry, Geneva University Hospital and University of Geneva, 24 Rue Micheli-du-Crest, CH-1211 Geneva 14 Switzerland. E-mail: denis.hochstrasser@dim.hcuge.ch

Manuel Peitsch
Geneva Biomedical Research Institute Glaxo, Wellcome Research and Development, 14 chemin des Aulx, Case Postale 674, CH-1228 Plan-les-Ouates Geneva Switzerland. E-mail: mcp13936@ggr.co.uk

Nicolle H. Packer
Australian Proteome Analysis Facility (APAF) and Macquarie University Centre for Analytical Biotechnology (MUCAB), School of Biological Sciences, Macquarie University, Sydney NSW 2109 Australia. E-mail: nicolle.packer@mq.edu.au

Vitaliano Pallini
Department of Molecular Biology, University of Siena, Pian dei Mantellini 44, I-53100 Siena Italy. E-mail: pallini@unisi.it

Jean-Charles Sanchez
Central Clinical Chemistry Laboratory, Geneva University Hospital, 24 rue Micheli-du-Crest, 1211 Geneva 14 Switzerland. E-mail: jean-charles.sanchez@dim.hcuge.ch

Marc R. Wilkins
Central Clinical Chemistry Laboratory and Department of Medical Biochemistry, Geneva University Hospital and University of Geneva, 24 rue Micheli-du-Crest, CH-1211 Geneva 14 Switzerland. E-mail: marc.wilkins@dim.hcuge.ch

Keith L. Williams
Australian Proteome Analysis Facility (APAF) and Macquarie University Centre for Analytical Biotechnology (MUCAB), School of Biological Sciences, Macquarie University, Sydney NSW 2109 Australia. E-mail: keith.williams@mq.edu.au

1 Introduction to the Proteome

Keith L.Williams and Denis F. Hochstrasser

1.1 Proteome: a new word, a new field of biology

This book has in its title a new word, "proteome". English is such a rich language that it is not often that new words are needed. So why have we chosen to make up this new word? We use proteome because something revolutionary is happening to biochemistry and the shock of a new word will help people to understand this profound change. Proteome indicates the PROTEins expressed by a genOME or tissue. Despite being first used in late 1994 at the Siena 2-D Electrophoresis meeting, the term proteome is already widely accepted. It is mentioned in more than 30 papers, including those from *Science* and *Nature*, and "Proteomics" has been the subject of a rash of conferences in 1997.

The antecedent for proteome is genome, itself a relatively new word (Friedrich 1996). Recently, genome has become a generic term for "big science" molecular biology. People think of grand dreams like the human genome project in the context of genome. And the genome projects have captured the imagination of scientific funding bodies and the biotechnology industry. By sequencing the entire genome of an organism, here for the first time in biology is the complexity of an organism understood at the level of information content.

Biology is full of unchartered territories, and some have compared the state of biology today to that of chemistry before the definition and filling in of the periodic table. Before this occurred there was a sense of the infinite, since the chemical world was made up of a seemingly limitless number of compounds. With the periodic table, defined by Dimitri Ivanovitch Mendeleiev in 1871, chemistry became finite. Born in 1834, this fourteenth child of a Russian family defined the table of elements by increasing atomic mass in March 1869 and revised it two years later, suggesting the existence of three new elements. He predicted their properties and indeed Gallium was discovered in 1878, Scandium in 1879 and Germanium in 1886. Following this there was the discovery of artificial elements. Ultimately the periodic table of elements comprised 83 natural and 19 artificial elements. While this did not change the huge complexity in the number of compounds known to chemistry, a simplicity was introduced because the ingredients making up those compounds were now understood.

How does the periodic table concept translate to biology? The most obvious analogy is with amino acids that are the building blocks of proteins. Such a biological periodic table would comprise up to perhaps 100 common "elements", being the 20 amino acids plus a large group of modified amino acids found in proteins (e.g. phosphoserine, phosphotyrosine, phosphothreonine, tyrosine sulfate, N-acetylglucosamine-linked to threonine or serine, N-acetylgalactosamine linked to threonine or serine, other O-linked sugars attached to threonine or serine, the numerous different asparagine-linked glycans, and a number of lesser known modified amino acids). The analogy to compounds in chemistry is then the very large number of combinations of amino acids that make up proteins.

But on the other hand, are genome studies and DNA sequencing projects to biology what the definition of the periodic table was to chemistry? This is a less convincing analogy, as DNA is a means of information storage in biology rather than the basic building block from which a living unit is made. Perhaps we can think of the genome as being the set of rules in each organism which dictate how the amino acid elements will be joined together to form proteins. By defining a genome, biology ceases to be unlimited, at least at the level of information. But we need also to define the proteome before we arrive at the ultimate definition of the core structural and functional molecules, through which other molecules (e.g. fats, carbohydrates) are synthesised in an organism.

The proteome, unlike the genome, is not a fixed feature of an organism. Instead, it changes with the state of development, the tissue or even the environmental conditions under which an organism finds itself. There are therefore many more proteins in a proteome than genes in a genome, especially for eukaryotes. This is because there can be various ways a gene is spliced in constructing mRNA, and there are many ways that the same protein can be post-translationally altered. So one of the famous dogmas of biology, the one-gene-one-enzyme hypothesis of Beadle and Tatum, is no longer tenable.

Returning to DNA, each organism has its own complement of genes which provide the information around which the organism is constructed. Some simple organisms already have their genomes fully sequenced and they contain from as few as 470 genes (for *Mycoplasma genitalium*) to 5885 genes (for *Saccharomyces cerevisiae*) (Table 1.1). The yeast *S. cerevisiae* is the only eukaryote genome sequenced as of writing this book. But progress on the human genome sequencing project is now hotting up, with a number of companies set up to do the slog of finding out the 3 billion or so combinations of the four bases adenine, guanine, cytosine, and thymine — the four letter code of DNA (Friedrich 1996). The sequence of all 70,000 to 100,000 or so human genes will probably be known within three to five years. This is an amazing feat, the pace of which has exceeded all expectations. Miklos and Rubin (1996) argue that the fundamental mammalian gene number may be as few as 12,000–14,000 genes, with the 6–8 fold increase in number being due to polyploidisation. This might make things simpler than one would have predicted at first sight, although presumably the polyploidisation allowed specialisation to occur, with the same basic gene developing different roles.

Table 1.1. Some organisms whose genomes are fully sequenced

Organism[a]	Genome size (Mb)	Number of genes[b]	Reference
Mycoplasma genitalium	0.58	470	Fraser et al. 1995
Methanococcus jannaschii	1.7	1,738	Bult et al. 1996
Haemophilus influenzae	1.8	1,743	Fleischmann et al. 1995
Synechocystis PCC6803	3.6	3,168	Kaneko et al. 1996
Escherichia coli	4.6	4,285	Blattner et al. 1997[c]
Saccharomyces cerevisiae	12.1	5,885	Goffeau et al. 1996

[a]Other organisms whose genome sequences will be available in the near future include *Bacillus subtilis*, *Staphylococcus aureus*, the insect *Drosophila melanogaster*, the plant *Arabidopsis thaliana*, and the nematode worm *Caenorhabditis elegans*.
[b]Predicted from the DNA sequence. Note that proteins smaller than approx. 10 kDa are often not predicted from genomic sequences, and so proteome studies are required to fill in the exact number of these small proteins that are actually made. This will in turn allow the relevant DNA sequences to be assigned as known genes.
[c]The complete sequence is available in EMBL/GenBank/DDBJ.

Whatever the truth is concerning gene number in mammals, industry is driving the discovery path, with a number of companies set up specifically to sequence DNA and use differential display of mRNA to search for key molecules involved in both normal physiological pathways and in disease. Some estimates suggest there are 30,000 known diseases. On the other hand, well-being is such an individual thing that one could argue that there are as many diseases as there are human beings! However, diseases for which major investment is likely to be commercially rewarded can be numbered in the hundreds rather than thousands (Drews 1996). Very few diseases are thought to have a simple genetic origin. Strohman (1994) estimates that only 2% of human disease results from a single gene defect (i.e. absence of protein product or presence of an altered protein), and that the genetic background of the individual often has a significant impact on the severity of the disease. Epigenetics, or how networks of genes are expressed, almost certainly has a major impact on complex diseases such as cancer. There appear to be a combination of several genetic alterations needed for cancer to occur. The short-cut to understanding these protein networks is to study them directly. This is proteomics.

Another powerful tactic to understand complexity is to combine molecular biology with advances in manipulation of whole organisms. Here one uses gene disruptions to try to understand how genes function (Miklos and Rubin 1996). It is interesting that one of the hard-core DNA sequencing companies recently acquired a small company which studies the biology of the nematode *Caenorhabditis ele-*

gans. The reasoning is that model organisms such as the yeast *Saccharomyces cerevisiae*, the slime mould *Dictyostelium discoideum*, the nematode worm *C. elegans*, the fruit fly *Drosophila melanogaster*, the plant *Arabidopsis thaliana*, and even the mouse *Mus musculus* can all have their genes disrupted. This is easiest in the simple organisms, but with major programs also being developed for the mouse, this process is even becoming routine in mammals. In theory, one can go directly from gene to function.

However, knocking out a single gene does not necessarily have straightforward consequences. Invariably the expression of many genes is altered in an organism even if only a single gene has been disrupted. This isn't surprising if one thinks of pathways of information flow, such that downstream of a blockage other things will be blocked, and upstream things may accumulate. Proteins that the deleted gene product normally interact with may be degraded in the absence of their normal partner.

A means of displaying and studying the products of genes directly is an attractive way of studying not only disease, but any complex problem in biology. Molecular understanding of how a cell operates in sickness and in health requires knowledge of the proteins and other cell components that are actually present, how they interact and the outcome of their interactions. The first step in such a major task is to find what parts of the genome are expressed, how much product is made, and how the products are modified. This book is concerned with explaining the technology which is being coordinated to achieve the goal of identifying, quantitating and studying the post-translational modifications to proteins in cells, tissues or even organisms. Such studies conducted in healthy and diseased tissues should, by differential analysis, give insight into what is altered. Proteomics is the direct approach.

The study of gene expression can also be attempted at the level of messenger RNA, and several ways of conducting such studies have recently been compared (Wan et al. 1996). While extremely powerful and now becoming automated to allow massive screening (Lockhart et al. 1996), it is important to understand that mRNA-based approaches measure message abundance rather than the actual proteins (the functional molecules). A protein cannot be synthesised without its mRNA being present, but you can have a protein in the cell when its mRNA is no longer present, and conversely you can have lots of mRNA and no translation of the message into protein. A recent study shows that there is not a good correlation between mRNA abundance and protein amount in a cell at a given time. Anderson and Seilhamer (1997) showed that in human liver mRNAs were enriched for secreted proteins, whilst mRNAs for cellular proteins were under-represented.

Thus one needs to screen the products of genes from cells. The problem is that until very recently protein science has been a slow and frustrating art. It has been quite common for the major efforts of a large research team to be taken up entirely with the purification and characterisation of a single protein. Unlike the developments in DNA sequencing and mRNA screening, where literally thousands of genes can be rapidly analysed in a well-equipped laboratory, this has not been possible for proteins so far. This book outlines the dramatic changes that have been converging in protein science and bioinformatics to produce the revolution that allows proteins, like DNA and mRNA, to be subject to mass screening.

This is proteomics and we hope that after reading this book you will have a better understanding of two things: i) the technology surrounding proteome analysis and ii) some insight into how proteome technology is already beginning to transform biology and medicine. You should also be able to envisage that one day biology will be fully documented at the proteome level, as it is currently becoming at the genome level.

1.2 The proteome and technology

In this introductory chapter we overview proteome technology and give a flavour of each of the following chapters. Some readers may not wish to read all of the more technical chapters, but the start of each chapter provides an overview of the key issues discussed. Hence if you want just a glimpse of the future, you might read the first and last three chapters and the introductions to those in between. We do hope however, that the material is sufficiently interesting that you will want to look into the undergrowth as well! Note that this book is not intended to be a laboratory manual, but instead a general and conceptual overview of this emerging and exciting field.

Protein technology is inherently more complex than DNA-based technology. Not only is the basic alphabet bigger (4 nucleotides for DNA, 20 unmodified and many more modified amino acids for proteins), but some genes can be variously spliced therefore making numerous different products from a single stretch of DNA. Additionally, mRNA editing is relatively common, leading to modified messages and corresponding protein products. There are also many ways in which proteins are modified after they have been synthesised. It can be argued with some justification that possibly all eukaryotic proteins are post-translationally modified in some way (e.g. truncation at the N- or C-terminus, by protein splicing (rarely), or by addition of various substituents such as sugars, phosphate, sulphate, methyl, acetyl or lipid groups). To make matters still more complicated, while an organism has effectively a single genome (if unusual circumstances such as the genes involved in antibody production in B- and T-cells are set aside), there are many proteomes. Even in an unicellular organism, the expressed proteins (proteome) will be different depending on the growth conditions.

The technologies required to separate large numbers of proteins, to identify them, and to study their modifications are by no means straightforward. As yet there is no equivalent to the Applied Biosystem 377 high-throughput DNA sequencer, which has served the genome projects so well. However, many technologies are converging and making proteome analysis possible. Our approach to pro-

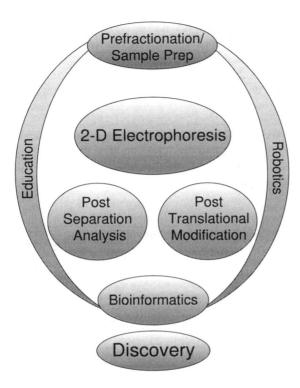

Fig. 1.1. The elements of proteome technology

teome studies has been to break down the technology into a number of components (Fig. 1.1), and the chapters of this book are arranged to follow the sequence of technologies from tissue fractionation, to large scale protein separation and display, to protein identification, to study of post-translational modifications, to the use of bioinformatics to unravel and systematise the information and ultimately use it for making predictions about protein structure and function. So this book is, in part, about the "picks and shovels" that will be the discovery vehicle in the coming proteome era of biology. Wrapped around these technologies are robotics to integrate and education to illuminate.

1.2.1 Thinking in two dimensions

A Medline search in April 1997 revealed 4,957 publications containing the words "two dimensional polyacrylamide gel electrophoresis". Hence there are many researchers who have papers involving this technology in their list of publications! But from whence did this technology evolve?

Electrophoresis methodology was invented almost half a century ago by Tiselius, and we could say that the two-dimensional (2-D) separation of proteins is over 40 years old. Separation of serum proteins with a combination of paper and starch gel electrophoresis was first done by Smithies and Poulik in 1956. Innovations in electrophoresis and the development of discontinuous buffer systems were rapidly applied to 2-D procedures. Gradient polyacrylamide gels were employed to enhance the sharpness of protein bands (Margolis and Kenrick 1967), and applied to 2-D separations (Margolis and Kenrick 1969). This paved the way for the application of isoelectric focusing (IEF) techniques to 2-D separations with either single concentration (Dale and Latner 1969; Macko and Stegemann 1969) or gradient (Emes et al. 1975) polyacrylamide gels. By 1975, a 2-D PAGE system had been developed which separated proteins on the basis of charge in the first separation dimension, using IEF, followed by size in the second separation dimension, using SDS-PAGE. The resolution of this method was immediately applied to the analysis of complex protein mixtures from whole cells or tissues (Iborra and Buhler 1976; Klose 1975; Scheele 1975). At about this time O'Farrell (1975) optimised the 2-D separation procedure that formed the basis for most subsequent developments in 2-D PAGE.

If all proteins in a cell or tissue are to be separated and displayed, it is crucial to be able to solubilise the proteins using techniques that are consistent with the separation process. Currently the only separation technology which provides highly purified proteins separated in a simple parallel process is 2-D electrophoresis. Other technologies, such as serial liquid chromatographic or capillary electrophoresis separations coupled to mass spectrometric detection, are being trialled. To our knowledge none of those systems yet provide reproducible and complete separation of complex protein mixtures.

This is not to say that 2-D electrophoretic separation of proteins is problem free. To date many studies with 2-D gel electrophoresis have involved the separation of soluble proteins. However, by solubilising proteins in suitable detergents, membrane proteins have also been studied. Recent work discussed in Chap. 2 shows that major advances have been made in displaying formerly intractable proteins using 2-D gel electrophoresis. A further issue often raised in relation to 2-D gel separations is finding the rare proteins in the midst of very common proteins. To find and characterise very rare proteins it will be essential to develop techniques for removing abundant proteins and concentrating the rare ones prior to the 2-D gel separation. Other issues concern contaminating cellular constituents such as lipids, DNA and salts, all of which can interfere with efficient and reliable 2-D gel separation. These issues are discussed in detail in Chap. 2.

In the eyes of many researchers not "in the field", 2-D polyacrylamide gel electrophoresis is still yet to become a routine separation vehicle. Why has it taken so long to become the method of choice for protein separation and display?

There are three reasons. Firstly, only recently have commercially manufactured immobilised pH gradients (IPGs) and precast SDS-PAGE gels of high quality become available. A basic requirement for reproducible gels is good quality control during the manufacture. Even within the same laboratory, it has been extraordinarily difficult to keep the same gel quality from day to day. Also the basis for the first dimensional separation is that proteins are focused electrophoretically in a pH gradient so that they move to a position in the gradient where they have no net charge (and so they stop moving). Traditionally the pH gradient has been established using mixtures of soluble charged compounds called carrier ampholytes. Since proteins are charged molecules, they can affect the pH gradient. This means that the exact amount of protein added to the first dimension separation itself can impact on the separation. This problem was solved by using immobilised pH gradients, where the compounds used to set up the pH gradient are chemically immobilised and so the gradient is stable. Use of IPGs solved the second problem which had limited progress in protein separation and analysis. Now much larger amounts of protein could be used in the separations so that chemical analysis of individual protein spots became possible. Thirdly, only now are computers and computer programs sufficiently inexpensive and powerful enough to be widely available and used for imaging 2-D gels.

1.2.2 Further dimensions in protein analysis

Having displayed the proteins on a 2-D gel, what is next? A 2-D gel is like stars in the sky — one has a map — but how do you determine which points are the stars and which the planets?

The holy grail of 2-D electrophoresis has always been to establish separations where individual protein spots can be identified and characterised. It is not widely appreciated that current 2-D methods have in many senses achieved this goal. Protein purification for the purpose of identification and characterisation, which was an impossibly slow and tedious process when purifying single molecules, is now solved for the majority of proteins (Chap. 2). The combination of learning how to load milligrams (or even tens of milligrams) of protein on a single 2-D gel with work on establishing high sensitivity characterisation techniques has brought about this revolution. Obviously the more protein loaded and separated on the 2-D gel, the less problem one has with very high sensitivity needed at the characterisation end. Ultimately, short of dressing up in space suits and building the sort of dust-free rooms used in microchip manufacture, one needs hundreds of femtomoles to low picomoles of protein on a single 2-D gel spot to be able to identify a protein and characterise it. To go lower in a conventional laboratory means that one spends a lot of time characterising human keratins, or less precisely, the proteins in the skin left behind by the experimentalists!

Protein identification in the genome era is easier than it was in earlier times. This is because there are now massive databases of inferred protein sequence established by DNA sequencing initiatives. If you work on *E. coli* or *S. cerevisiae* then all the genes encoding the proteins have been sequenced. So protein identification can be approached as a pattern matching exercise, rather than a protein sequencing exercise. Many attributes can be used. These include apparent mass and isoelectric point (estimated from the 2-D gel), N-or C-terminal "tag sequences", amino acid composition and peptide mass data. These are discussed in Chap. 3.

It is interesting how quickly the emphasis has gone from a focus on protein identification to characterisation. Already we are not content just to know that a protein is, for example, glutamate dehydrogenase. Now we want to know how much of the protein is present in the cell, and if it starts (its N-terminus) and ends (its C-terminus) where the molecular biologists predicted it should. With eukaryotes we often find that it doesn't! Also we would like an accurate estimate of the mass of the protein, as this will indicate if the protein is likely to be post-translationally modified. If we think that a protein has modifications, we might want to further characterise the protein and do this on a single protein spot from a 2-D gel. Perhaps we would like to be able to profile the N-linked oligosaccharides, do a monosaccharide profile to get clues about the O-linked sugars and finally estimate the amounts of phosphoamino acids (phospho-serine, -threonine and -tyrosine). All of this is becoming possible and is described in Chap. 4. Of course for complete analysis of post-translational modifications, such as assigning the specific sites of glycosylation, more detailed studies are needed. But this too is becoming possible on picomole amounts of protein.

1.2.3 Information and the proteome

So there is a revolution in protein separation, display, and protein characterisation. The next step is to develop technologies to handle quite massive datasets. With genomics the problems are relatively straightforward. One must catalogue and make accessible huge strings of a four letter alphabet. Of course there are major challenges to be able to compare or piece together large amounts of sequences, but fundamentally the information is linear and it is organism specific. Proteomics is different. First, every cell type or tissue has its own proteome, and this represents only part of the genome. To make matters more complicated, it is often the case that the way a gene is expressed in one cell type or tissue is different from the way it is expressed in another, and the same gene product can come in different forms even in a single cell type! The most obvious way that proteins change within and between cell types and tissues involves glycosylation, but this is only one of many alterations. Many times these different forms are immediately apparent as spots occupying a different map position on the 2-D gel.

All of the above is an intriguing challenge for the brave bioinformatician. How to best collect, systematise and make readily accessible all of this information can only be glimpsed at this stage, but in Chap. 5 and 6 we sketch the new proteome bioinformatics world. This includes how we move textual or image proteomic data around different types of databases, and thus ask such things as: how does a disease tissue compare with healthy tissue? Being able to ask such questions will enormously increase our understanding of biology and move us towards the understanding of function.

Ultimately we would like to be able to progress from protein sequence to three-dimensional structure. Is this a realistic goal? There is still much to be learned, especially concerning how post-translational modification affects protein shape. But the huge developments in structural biology and computer modelling are now giving insights as to what the structures of very large numbers of proteins, including ones never previously crystallised, might be. Here again the integration of databases is crucial so that one can move from a spot on a 2-D gel, to its sequence in a DNA database, to annotation as to its post-translational modifications, to tools that can show or predict a three dimensional structure. The art of structure prediction is addressed in Chap. 7.

The above discussion considers proteins as individuals. Will we eventually have information about the proteins with which a particular protein can interact? It is challenging to think about how higher order databases may be constructed.

1.3 Looking towards new frontiers

Much of the above is technofix stuff, but you can't discover things if you don't have the tools. We do not go beyond discussion of the individual components of proteome instrumentation here. Obviously the next step is to integrate a suite of instruments if one wishes to begin large scale screening. It is important to understand that a major reason for developing robotics is to establish the highest possible quality control standards. But this is a matter for another book.

Why are researchers bothering to develop all of these new methods? Clearly it is because they see new horizons and hope this will enable a greater understanding of biology. Two chapters (Chap. 8 and 9) give a glimpse of the kinds of discovery that proteomics is allowing. These chapters are deliberately visionary to try to sketch out where the field is going. Of course, as with all new developments, it is likely that some of the things we see as important won't be. But we are also confident that many as yet unanticipated discoveries will be made.

So with the guideposts that we have provided, please read on and enjoy the exploration of the proteome.

Acknowledgments

We would like to thank Neil Barclay (MRC Cellular Immunology Unit, University of Oxford, UK) and Marc Wilkins for helpful comments on this chapter, and Brad Walsh (APAF) who prepared Fig. 1.1.

References

Anderson L, Seilhamer J (1997) A comparison of selected mRNA and protein abundances in human liver. Electrophoresis 18:533–537

Blattner FR, Plunkett G III, Mayhew GF, Perna NT, Glasner FD (1997) Submitted to the EMBL/GenBank/DDBJ database

Bult CJ, White O, Olsen GJ, Zhou L, Fleischmann RD, Sutton GG, Blake JA, Fitzgerald LM, Clayton RA, Gocayne JD, Kerlavage AR, Dougherty BA, Tomb JF, Adams MD, Reich CI, Overbeek R, Kirkness EF, Weinstock KG, Merrick JM, Glodek A, Scott JL, Geoghagen NSM, Weidman JF, Fuhrmann JL, Nguyen D, Utterback TR, Kelly JM, Peterson JD, Sadow PW, Hanna MC, Cotton MD, Roberts KM, Hurst MA, Kaine BP, Borodovsky M, Klenk HP, Fraser CM, Smith HO, Venter JC (1996) Complete genome sequence of the methanogenic archaeon, *Methanococcus jannaschii*. Science 273:1058–1073

Dale G, Latner AL (1969) Isoelectric focusing of serum proteins in acrylamide gels followed by electrophoresis. Clin Chim Acta 24:61–68

Drews J (1996) Genomic sciences and the medicine of tomorrow. Nat Biotechnol 14:1516–1518

Emes AV, Latner AL, Martin JA (1975) Electrofocusing followed by gradient electrophoresis: a two-dimensional polyacrylamide gel technique for the separation of proteins and its application to the immunoglobulins. Clin Chim Acta 64:69–78

Fleischmann RD, Adams MD, White O, Clayton RA, Kirkness EF, Kerlavage AR, Bult CJ, Tomb JF, Dougherty BA, Merrick JM, McKenney K, Sutton G, Fitzhugh W, Fields C, Gocayne JD, Scott J, Shirley R, Liu LI, Glodek A, Kelley JM, Weidman JF, Phillips CA, Spriggs T, Hedblom E, Cotton MD, Utterback TR, Hanna MC, Nguyen DT, Saudek DM, Brandon RC, Fine LD, Fritchman JL, Fuhrmann JL, Geoghagen NSM, Gnehm CL, McDonald LA, Small KV, Fraser CM, Smith HO, Venter JC (1995) Whole-genome random sequencing and assembly of *Haemophilus influenzae* Rd. Science 269:496–512

Fraser CM, Gocayne JD, White O, Adams MD, Clayton RA, Fleischmann RD, Bult CJ, Kerlavage AR, Sutton G, Kelley JM Fritchmann JL, Weidman JF, Small KV, Sandusky M, Fuhrmann J, Nguyen D, Utterback TR, Saudek DM, Phillips CA, Merrick JM, Tomb JF, Dougherty BA, Bott KF, Hu PC, Lucier TS, Peterson SN, Smith HO, Hutchinson CA, Venter JC (1995) The minimal gene complement of *Mycoplasma genitalium*. Science 270:397–403

Friedrich GA (1996) Moving beyond the genome projects. Nat Biotechnol 14:1234–1237

Goffeau A, Barrell BG, Bussey H, Davis RW, Dujon B, Feldmann H, Galibert F, Hoheisel JD, Jacq C, Johnston M, Louis EJ, Mewes HW, Murakami Y, Philippsen P, Tettelin H, Oliver SG (1996) Life with 6000 genes. Science 274:546–567

Iborra F, Buhler JM (1976) Protein subunit mapping. A sensitive high resolution method. Anal Biochem 74:503–511

Kaneko T, Sato S, Kotani H, Tanaka A, Asamizu E, Nakamura Y, Miyajima N, Hirosawa M, Sugiura M, Sasamoto S, Kimura T, Hosouchi T, Matsuno A, Muraki A, Nakazaki N, Naruo K, Okumura S, Shimpo S, Takeuchi C, Wada T, Watanabe A, Yamada M, Yasuda M, Tabata S (1996) Sequence analysis of the genome of the unicellular cyanobacterium Synechocystis sp. strain PCC6803. II. Sequence determination of the entire genome and assignment of potential protein-coding regions. DNA Res 3:109–136

Klose J (1975) Protein mapping by combined isoelectric focusing and electrophoresis in mouse tissues. A novel approach to testing for induced point mutations in mammals. Humangenetik 26:231–243

Lockhart DJ, Dong H, Byrne MC, Follettie MT, Gallo MV, Chee MS, Mittmann M, Wang C, Kobayashi M, Horton H, Brown EL (1996) Expression monitoring by hybridization to high-density oligonucleotide arrays. Nat Biotechnol 14:1675–1680

Macko V, Stegemann H (1969) Mapping of potato proteins by combined electrofocusing and electrophoresis identification of varieties. Hoppe Seylers Z Physiol Chem 350:917–919

Margolis J, Kenrick KG (1967) Polyacrylamide gel-electrophoresis across a molecular sieve gradient. Nature 214:1334–1336

Margolis J, Kenrick KG (1969) 2-dimensional resolution of plasma proteins by combination of polyacrylamide disc and gradient gel electrophoresis. Nature 221:1056–1057

Miklos GL, Rubin GM (1996) The role of the genome project in determining gene function: insights from model organisms. Cell 86:521–529

O'Farrell PH (1975) High resolution separation two-dimensional electrophoresis of proteins. J Biol Chem 250:4007–4021

Scheele GA (1975) Two-dimensional gel analysis of soluble proteins. Characterisation of guinea pig exocrine pancreatic proteins. J Biol Chem 250:5375–5385

Smithies O, Poulik MD (1956) Two-dimensional electrophoresis of serum proteins. Nature 176:1256–1266

Strohman R (1994) Epigenesis: the missing beat in biotechnology. Bio/Technology 12:156–164

Wan JS, Sharp SJ, Poirer GM, Wagaman PC, Chambers J, Pyati J, Hom YL, Galindo JE, Huvar A, Peterson PA, Jackson MR, Erlander MG (1996) Cloning differentially expressed mRNAs. Nat Biotechnol 14:1685–1691

2 Two-Dimensional Electrophoresis: The State of the Art and Future Directions

Ben R. Herbert, Jean-Charles Sanchez and Luca Bini

2.1 Introduction

Two-dimensional polyacrylamide gel electrophoresis (2-D PAGE) is the only method currently available which is capable of simultaneously separating thousands of proteins. Thus 2-D PAGE is the heart of proteome technology. The first dimension of 2-D PAGE is isoelectric focusing (IEF), during which proteins are separated in a pH gradient until they reach a stationary position where their net charge is zero. The pH at which a protein has zero net charge is called its isoelectric point (pI). In the second dimension the proteins separated by IEF are separated orthogonally by electrophoresis in the presence of sodium dodecyl sulphate (SDS-PAGE). The surfactant SDS binds to proteins, overriding their intrinsic charge, such that they all have the same charge density and free solution electrophoretic mobility. When the SDS-coated proteins migrate in a sieving polyacrylamide gel they separate based on their molecular mass. The high resolution of 2-D PAGE results from the first and second dimension separations being based on independent protein parameters.

In the early 1970s the pioneers of 2-D PAGE immediately recognised the potential of the technique for studies that would now be classified as proteome projects. When Kenrick and Margolis (1970) combined native isoelectric focusing (IEF) with pore gradient SDS-PAGE to separate serum proteins, the 2-D PAGE revolution had begun. The authors commented in their conclusions that 2-D PAGE:

"has obvious application in the typing of genetically polymorphic material, in the testing of protein heterogeneity, and in the resolution of complex mixtures of proteins".

Five years later, three papers (Klose 1975; O'Farrell 1975; Scheele 1975) described 2-D PAGE using denaturing IEF, in a form that most users would be familiar with to this day. The resolving power of the technique was amply demonstrated in O'Farrell's separation of over 1000 *Escherichia coli* proteins, and O'Farrell discussed the use of 2-D PAGE to investigate charge heterogeneity arising from post-translational modifications such as phosphorylation, acetylation and glycosylation. The study of post-translational modifications and their influence on protein function is now a major focus of many proteome projects and Chap. 4 is dedicated to this topic.

The potential of 2-D PAGE was not immediately fulfilled, mainly because of drawbacks with the carrier ampholytes used to generate pH gradients. Using carrier ampholytes it was difficult to achieve good reproducibility within a laboratory, and inter-laboratory comparisons were virtually impossible. Thus data sharing between laboratories was not practical. A second and equally frustrating limitation of ampholytes was the inability to separate preparative loads of protein, which often necessitated running multiple gels and pooling the desired protein species for analysis. Thus it was often not possible to proceed beyond a 'stars in the sky' approach to 2-D PAGE where the maps produced were of little use.

Proteome analysis became feasible when analysis of single 2-D PAGE spots from a single gel became a reality (see Chap. 3 and 4). The introduction of immobilised pH gradient (IPG) gels eliminated the problems of gradient instability and poor sample loading capacity associated with carrier ampholyte pH gradients (Bjellqvist et al. 1982; Görg et al. 1988; Righetti 1990). To form an IPG the pH gradient is co-polymerised with the acrylamide gel matrix, which results in completely stable gradients at all but the most alkaline (> 12) pH values. The availability of commercial precast IPG gels in a variety of narrow and broad pH ranges has brought 2-D PAGE out of the realm of the dedicated specialist and to the forefront of high resolution protein separation. Using the commercially available IPGs as the first dimension of 2-D PAGE it is possible to create highly reproducible reference maps, as well as separate milligram quantities of protein for micropreparative purposes (Bjellqvist et al. 1993a,b; Hanash et al. 1991). With the aid of dedicated computer software, maps produced by different laboratories can be directly compared (Corbett et al. 1994; Blomberg et al. 1995). The reproducibility of 2-D PAGE reference maps using IPG gels has prompted several groups to make their maps available over the World-Wide Web. Some examples include the SWISS-2DPAGE[1] and HSC-2DPAGE[2] databases (see Chap. 6).

As with every quantum development, a step forward opens many avenues previously blocked. Here we address some areas where major progress is being made. Sample preparation and solubilisation is a key area of 2-D PAGE needing further work, especially for insoluble samples such as membrane and nuclear proteins, or proteins from highly resistant tissues like hair and skin. Low abundance proteins often have important regulatory roles, but many will remain undetected on broad pH range micropreparative 2-D PAGE gels. Prefractionation and/or very high (> 50 mg) sample loads on IPG gels will be necessary to separate micropreparative amounts of low abundance proteins from complex mixtures. High sample loads, which provide approximately 50 picomoles of each separated protein, are required for analysis of co-translational and post-translational modifications (see Chap. 4).

---

[1] http://www.expasy.ch/ch2d/ch2d-top.html

[2] http://www.harefield.nthames.nhs.uk/

Separation of basic (pI > 10) proteins on IPGs has not been possible until recently because of the instability of the gel matrix at high pH and problems with protein streaking caused by electroendosmotic flow. These issues are being addressed and separation up to pH 12 is now possible (Görg et al. 1997). Further improvements are still likely to be made. In addition to the challenges of sample solubilisation and separation, the quantitation of proteins on 2-D PAGE maps is another technical problem which must be solved to allow the tracking of protein expression changes during development in different tissues or in disease.

2.2 Sample preparation

Pretreatment of samples for IEF involves solubilisation, denaturation and reduction to completely break the interactions between proteins and to remove non-protein sample components such as nucleic acids (Rabilloud 1996). Ideally one would achieve complete sample solubilisation in a single step and thus avoid unnecessary handling, as is the case for soluble samples which can be readily taken up in the typical IEF sample solution of 8 M urea, 4% 3-[(3-cholamidopropyl)dimethylam-monio]-1-propanesulfonate (CHAPS), 50–100 mM dithiothreitol (DTT) and 40 mM Tris. The challenge for 2-D PAGE is the solubilisation and separation of insoluble samples such as membrane-associated proteins or those from tissues that are highly resistant to denaturation.

In an ideal world samples could be applied to 2-D gels unfractionated, but this is not always practical especially with very complex mixtures or when abundant proteins dominate the sample as in the case of albumin in plasma. Organelle and/or plasma membrane fractions can be used to considerably reduce the complexity of cellular samples (Rabilloud et al. 1997). In addition, a range of preparative electrophoretic techniques have been used to fractionate samples prior to 2-D PAGE, and these techniques are discussed in Sect. 2.4.1.

2.2.1 Increasing protein solubility with chaotropes and surfactants

Many samples, such as wool proteins and membrane proteins, are solubilised in the standard sample solution only to become insoluble during the course of the prolonged focusing required during the first dimension IPG separation. Rabilloud et al. (1997) reported the selective adsorption of hydrophobic membrane proteins to the IPG during the IEF, resulting in these proteins being under-represented on the second dimension gel. The incorporation of up to 2 M thiourea and mixtures of CHAPS and other sulphobetaine surfactants improved the solubilisation of the proteins during the IEF, increasing the transfer of proteins to the second dimension gel. This suggests that hydrophobic interactions between the proteins and the IPG gel matrix are responsible for the losses. Other recent studies have shown that some of the acrylamido buffers used to form IPGs are hydrophobic, especially the basic buffers, which leads to hydrophobic interactions between the gel matrix and proteins from the sample (Esteve-Romero et al. 1996; Rabilloud 1994).

2.2.2 The choice of reducing agent

Aside from the issue of protein losses due to hydrophobic interactions, a separate problem involves the choice of reducing agents used in sample solutions for IEF. In order to completely solubilise complex mixtures of proteins for IEF, it is necessary to completely break inter- and intra-chain disulphide bonds by reduction. This is usually achieved with a free-thiol containing reducing agent such as β-mercaptoethanol or DTT (Rabilloud 1996). However reagents such as DTT are charged and thus migrate out of the pH gradient during the IEF, which results in a loss of solubility for some proteins, especially those which are prone to interaction by disulphide bonding such as keratins and keratin associated proteins from hair and wool. Replacing the thiol containing reducing agents with a non-charged reducing agent such as tributyl phosphine (TBP) greatly enhances protein solubility during the IEF and results in increased transfer to the second dimension (Herbert et al. 1997).

Thiourea and TBP increase protein solubility through different routes, hence they can be seen as complementary reagents. Thiourea increases the chaotropic power of the sample solution and TBP ensures complete reducing conditions during the IEF. Thus it is advantageous to combine the two in a modified sample solution (Rabilloud et al. 1997; Herbert et al. 1997). Fig. 2.1 shows proteins from *Dictyostelium discoideum* cells separated using TBP during the IEF. Not only are more proteins displayed than with conventional techniques, but the focusing time is reduced and the protein spots are more sharply focused.

2.2.3 Removal of nucleic acids

The presence of nucleic acids, especially DNA, has a severe detrimental effect on the separation of proteins by IEF for a number of reasons. Under denaturing conditions, such as in the sample solutions described above, DNA complexes are dissociated and cause a marked increase in the viscosity of the solution. This inhibits protein entry and slows migration in the IPG gel. In addition, DNA binds to proteins in the sample and causes artifactual migration and streaking (Rabilloud 1996).

There are two methods of nucleic acid removal which are suitable for maintaining sample integrity for IEF. The first is to utilise the ability of carrier ampholytes to form complexes with nucleic acids and then remove the complexes using ultracentrifugation. If the extraction is done at high pH, proteins behave as anions and binding to the anionic nucleic acids is minimised (Rabilloud et al. 1986). The second method is enzymatic digestion, which is achieved by adding pure endonucle-

pH 3.5   **10**

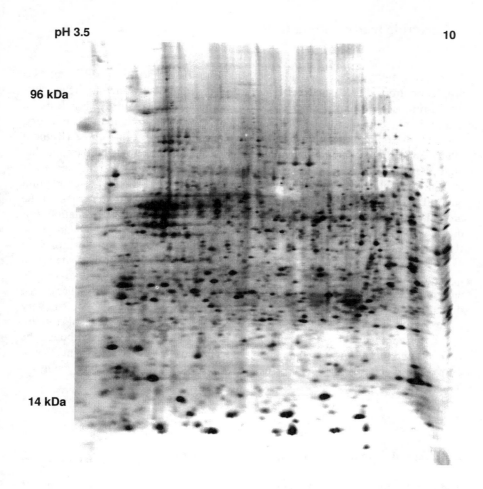

96 kDa

14 kDa

Fig. 2.1. Silver stained 2-D PAGE gel of the multicellular aggregate (slug) stage of the slime mould *Dictyostelium discoideum*. 250 μg of lyophilised slugs were solubilised in 500 μl of 7 M urea, 2 M thiourea, 4% CHAPS, 2 mM TBP, 40 mM Tris and 0.5% carrier ampholytes 3–10. The pH 3.5–10 non-linear IPG was rehydrated with the 500 μl of sample and IEF was performed for 2.5 hour at 300 V, 1 hour at 1,000 V, 1 hour at 2,500 V and a final phase of 5,000 V, to give a total volthour product of 90 kVh. The second dimension was a 8–18% Bio-Rad Protean IIxi SDS-PAGE gradient gel of thickness 1.5 mm. The finished gel was stained using a diamine silver staining protocol (Hochstrasser and Merril 1988)

ase to the sample after solubilisation in sample solution at high pH (40 mM Tris), thus minimising the action of contaminating proteases in the sample. The advantage of the endonuclease method is that sample preparation can be achieved in a single step, by the addition of the enzyme prior to loading on to the IPG gel.

2.3 Sample loading on IPG gels

There are many ways of applying the sample to the surface of an IPG gel: as a droplet or streak directly on the gel surface, in a depression formed in the gel during casting, soaked into a paper strip or in a sample cup. Because IEF is truly an equilibrium technique, the actual application point does not matter in theory, although the optimal position will be governed by the type of sample (Righetti 1990). For example, samples containing SDS will be better applied at the cathodic, high pH, end of the IPG to allow this highly charged surfactant to migrate away from the proteins with a minimum of disruption. When loading at a discrete point on an IPG it is usually better to load at one of the pH extremes, where most of the proteins will be charged and thus minimise sample loss through precipitation. For this reason, until recently high loading was achieved with a cup at one end of the IPG strip. However, it is very difficult, even with the most soluble samples, to avoid some sample loss at the loading point and this problem is exacerbated with high protein loads (Bjellqvist et al. 1993b). To increase the amount of protein entering the gel when loading with cups, it is necessary to radically change the dimensions of the IPG, especially to widen the loading area to accommodate very large cups, and use long sample entry times at low voltage (Bjellqvist et al. 1993b). Using this approach it is possible to load as much as 15 mg of protein, although the IPG gels and sample cups are non-standard which does not lend itself to inter-laboratory reproducibility. New techniques as described below are revolutionising gel loading.

2.3.1 Sample application during IPG rehydration

The fact that the commercial IPG gels are supplied dehydrated and that sample can be applied at any position on an IPG allows the sample to be loaded during the gel rehydration, over the whole of the IPG, thus avoiding sample loss by precipitation in sample cups (Rabilloud et al. 1994; Sanchez et al. 1997a). This method of sample application offers many advantages over cup loading and is rapidly becoming the loading method of choice. The elimination of sample loss provides a means of quantitative sample loading for analytical purposes and very high (> 10 mg) micro-preparative loads can be achieved on standard commercial IPG gels. The long sample entry phases used with cup loading can be reduced to less than 5 hours with sample loading by rehydration, and the focusing time can be reduced even further from 3 to 5 days to less than 1 day by combining the solubilising power of thiourea and TBP with rehydration loading. Use of TBP and rehydration during loading allows even highly insoluble samples such as wool proteins to be focused in less than 80 kVh which is only an overnight run when using a 5,000 V power supply (data not shown). Sanchez et al. (1997a) have developed a simplified apparatus for rehydration loading which accommodates up to 10 IPG gels in separate grooves, to eliminate sample cross-contamination.

2.4 Low abundance proteins

As discussed in the previous sections, 2-D PAGE plays a key role in the proteome approach. It is currently the most powerful technique for the simultaneous study of protein expression and post-translational modification. However, not all the proteins expressed by a genome are detectable on a single 2-D gel with current methodologies. One thing which still remains unclear is what copy number we need before a protein can be detected on a gel.

Fig. 2.2 considers the relationship between the quantity of protein loaded onto a 2-D gel and the final concentration of a 20 kDa protein present at 10, 10^3 and 10^5 copies per cell. Assuming that no prefractionation is performed, with the loading capacity of 10 mg per gel only the proteins present at more than 10^5 copies per cell (17 pmol) can be easily identified using analytical procedures. With a highly sensitive silver stain, proteins present at more than 10^3 copies per cell (0.17 pmol) will be detectable, but difficult to analyse with analytical techniques. Immunoblotting using a combination of high affinity monoclonal antibodies and enhanced chemiluminescence (ECL) will allow detection of proteins present at more than 10 copies per cell (1.7 fmol). Therefore, the proportion of the total protein complement that

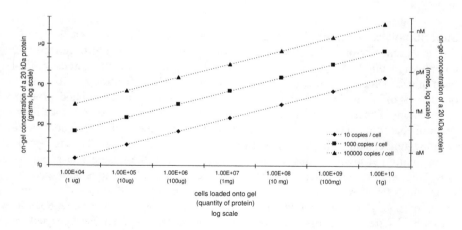

Fig. 2.2. Final concentration of a 20 kDa protein after 2-D gel electrophoresis and how it varies with the quantity of cells of proteins loaded onto the gel, and the copy number of the protein per cell (Wilkins et al. unpublished). This is based on 10^9 cells being 1 g fresh weight, 200 mg dry weight and containing 100 mg protein

can be seen on a gel for any cell or tissue will depend on the copy number, on the quantity of protein loaded on the gel and on the method of detection. Enrichment of specific subcellular fractions prior to the loading of the sample on the gel will increase the number of detectable low abundance proteins. For relatively small genomes such as *Saccharomyces cerevisiae* (5,885 genes) or *E. coli* (4,285 genes), 40 to 50% of their proteome can be displayed on a 160 × 180 × 1.5 mm 2-D PAGE gel. On the same type of gel, not more that 20% of the expressed genes of mammalian cells (5,000 proteins corresponding probably to 10,000 to 15,000 polypeptides or isoforms) are detectable, representing only the tip of the proteome "iceberg" in this case (Celis et al. 1992). Since many of the low-copy number proteins are likely to have very important regulatory functions in cells, it is clear that the separation of low copy number proteins in amounts sufficient for post-separation analysis is an important issue in proteome studies and presents a challenge for 2-D techniques. Below we describe some ideas on how to prefractionate the sample prior to its 2-D separation, increase the loading capacity of IPG gels, and enhance the protein detection sensitivity.

2.4.1 Cell fractionation and protein prefractionation

Despite the high sensitivity of several detection methodologies including infrared matrix assisted laser desorption / ionisation mass spectrometry (IR-MALDI-MS) or multiple immuno 2-D blotting, the analysis of low abundance proteins remains challenging. Two alternative approaches of enrichment of rare polypeptides prior to 2-D PAGE separation are described here: subcellular fractionation and protein prefractionation.

Subcellular fractionation. It is possible to apply traditional biochemical separation techniques prior to 2-D PAGE separation, with the goal being to reduce the complexity from thousands of proteins in the whole cell sample to hundreds of proteins in each subcellular fraction. Cell fractionation and purification of subcellular organelles offers great potential to obtain highly purified and enriched protein preparations. The most common procedures are those for preparation of subcellular fractions based on size or density differences. The first step of these methods is to separate nuclei and unbroken cells from cytoplasmic organelles by differential sedimentation at low centrifugal force in order to obtain the postnuclear supernatant. This supernatant is then subject to various density gradients to isolate specific organelles such as mitochondria or lysosomes. Huber et al. (1996) investigated the effect of oncogenic transformation on the molecular organisation of endocytic organelles. After subcellular fractionation, light membrane fractions enriched in endosomes were applied to 2-D PAGE. Computer-assisted image analysis revealed differential protein expression in K-ras-transformed cells versus control cells. Another approach based on the charge properties of organelles is also possible. The equipment that achieves this organelle separation is called free-flow electrophoresis.

Electrophoretic fractionation of proteins. Preparative isoelectric focusing as a protein prefractionation procedure was proposed by Bier et al. (1979) and a commercial version called the Rotofor has been produced by Bio-Rad (Hercules, CA). This method fractionates proteins into defined pH ranges by liquid-phase isoelectric focusing and is capable of 500-fold purifications of proteins from complex mixtures. Clearly such two-log concentrations have a major impact on separating and characterising rare proteins. It has been built as a rotating chamber divided into 20 compartments, into which the sample, diluted in 2% ampholytes, is loaded. Once the pH gradient has been established, the proteins migrate to their respective isoelectric points and are thus concentrated. At the end of the experiment, the different pH fractions are obtained from the unit. The Rotofor can be used under mild conditions or under strongly denaturing conditions by adding chaotropes, surfactants and reducing agents. More than one gram of plasma proteins have been efficiently separated with this procedure and used for subsequent 2-D PAGE analysis (Sanchez et al. 1992). Immobilised pH gradient technology has also been incorporated into a preparative IEF apparatus. Righetti et al. (1989) described a multicompartment electrolyser where each compartment is separated by a polyacrylamide gel membrane with a specific pH. A commercial version of this apparatus, called the IsoPrime, is marketed by Hoefer Pharmacia (San Francisco, CA). Acrylamido buffers are incorporated into the polyacrylamide membranes in the same way that they are used to create pH gradients in IEF gels. Liquid-phase IEF is conducted, usually under denaturing conditions, with the main difference from the Rotofor being that the pH specificity of the membranes allows very defined pH gradients to be constructed. The majority of applications of this apparatus have been in purifying a single protein from a mixture, and its utility for protein fractionation for 2-D PAGE has not been thoroughly investigated. Another promising electrokinetic apparatus, called the Gradiflow, has been described by Corthals et al. (1997). Samples are separated by charge and/or size under mild electrophoretic conditions without denaturing additives such as SDS or reducing agents, although the system can also be used under denaturing conditions. This technique represents a means for enrichment of low abundance proteins in the near future.

If a preparative IEF fractionation has been performed, it is sensible to use narrow pH range IPG gels for the subsequent 2-D PAGE separation. It is now becoming apparent that narrow pH range IPG gels can be used with high loads of unfractionated samples, eliminating the need for a prefractionation. Because of the simplicity of this procedure, it has considerable promise for sample fractionation as is described below.

2.4.2 High protein loads combined with narrow pH gradients

A simple way to increase the number of detectable spots on a 2-D gel is to improve the IPG loading capacity. Righetti (1990) studied and optimised the three major parameters that could improve this capacity by increasing i) the ionic strength of the gel from 2 to 10 mequiv./l, which allowed a 4-fold increment in loading capacity, ii) the gel thickness from 1 to 5 mm, which gave a 5-fold improvement in loading capacity, and finally iii) by reducing the width of the pH interval which resulted in a linear increase in loading capacity. The dimensions of commercially available IPG gels (Pharmacia) are fixed at 0.5 mm thick and 3 mm wide, and only pH 3.5–10 and pH 4–7 gradients are currently available as pre-cut gel strips. Hanash et al. (1991) applied increasingly concentrated samples to a commercially available 3 mm wide sigmoidal pH 3.5–10 IPG gel. However, a protein content above 1.5 mg resulted in protein aggregation and precipitation at the application point. To apply higher protein loads it was necessary to increase the sample volume beyond the 100 µl maximum of the Pharmacia sample cups. As already mentioned in Sect. 2.3, Rabilloud et al. (1994) and more recently Sanchez et al. (1997a) demonstrated that it was possible to increase the sample volume to 500 µl using in-gel rehydration and thus load up to 5 mg of protein samples on wide pH range IPG gels and up to 15 mg of some samples on narrow pH range gels. In this approach, the entire IPG gel is used for sample application, with the protein entering the gel during its rehydration (e.g. Fig. 2.3). In the near future we might be able to load up to 120 mg of protein on a single 2-D gel by increasing the 18 cm IPG strip thickness from 0.5 to 2 mm and its width from 3 to 6 mm (Sanchez et al. 1997a). Therefore, electrophoretic or chromatographic prefractionation techniques, followed by high loading of narrow-range IPG strips by rehydration should make possible the study of very rare proteins in complex samples, and the identification of the majority of spots on 2-D PAGE reference maps from any proteome.

2.4.3 Sensitive detection

Immunodetection is a specific and sensitive technique to identify single protein spots from 2-D PAGE, and it can detect as little as tens of femtomoles amounts of protein, depending on the specificity of the antibodies. With this method, Celis et al. (1995) identified and mapped components of signal transduction pathways including ras, MEK2, Cip1, cdk2, ERK and RhoA. However, it is a slow technique, as it is only possible to identify a few proteins per gel per day. Another disadvantage with this approach is that detection only requires the epitope recognised by the antibody, and thus no information on protein truncation or modification can be obtained unless the 2-D map position of the unmodified molecule is known. Recently, a multiple immuno 2-D blotting (MI-2DB) method using several monoclonal antibodies was developed to simultaneously analyse oncogene expression and cell cycle checkpoints in patient solid tumour biopsies and transformed cell lines (Sanchez et al. 1997b; see Sect. 8.7.6). The antibody mixture detected nine low abundance proteins simultaneously and interestingly, when antibodies were pooled, the sensitivity of detection was increased. These preliminary results demonstrate that MI-2DB is an excellent technique to display, quantify and simultaneously measure the expression and indicate potential post-translational

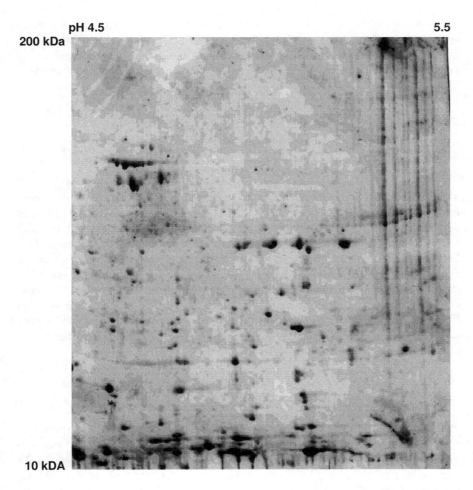

Fig. 2.3. Amido-black stained PVDF membrane of *Escherichia coli* proteins. Four mg of whole cell lysate was loaded onto a narrow pH range (4.5–5.5) IPG strip by in-gel rehydration. Proteins were then focused, separated on a second dimension SDS gel and transferred to a PVDF membrane according to Sanchez et al. (1997a). Note that each visible protein is present in micropreparative quantities

modifications of these low copy number gene products. However, this is only possible provided that the proteins of interest are well separated on the gel and the antibodies do not recognise epitopes which may change, such as protein glycosylation.

A novel approach has been recently described by Eckerskorn et al. (1997) for the sensitive analysis and identification of proteins. Plasma proteins were separated by 2-D PAGE, transferred onto a PVDF membrane, incubated with succinic acid and scanned with IR-MALDI-MS. They demonstrated that the sensitivity for protein detection was better than that of sensitive silver stained gels. In addition, they were able to propose the presence of post-translational modifications by comparing the measured mass to the theoretical mass deduced from the DNA sequence (see Sect. 3.3.3).

2.5 Basic proteins

The separation of proteins with pI values above pH 10, such as histones and ribosomal proteins, has always been a difficult problem for IEF because at the extremes of pH the buffering power of water increases dramatically and causes high conductivity (Righetti 1990). The system which has been the most widely used for IEF of basic proteins is nonequilibrium pH gradient electrophoresis (NEPHGE) where the proteins are focused to allow a separation to occur but the pattern never reaches a steady state. If the focusing is prolonged to attempt to reach equilibrium, the separation is lost as the gradient breaks down. Consequently NEPHGE gives poor resolution and reproducibility is difficult to achieve because the separation is very sensitive to parameters such as sample composition, run time, gel length and ampholyte choice (Dunn 1993).

2.5.1 Separation of basic proteins on IPG gels

In theory IPG gels have the ability to provide an equilibrium focusing pattern of very basic proteins, although until recently there was no suitable strongly basic acrylamido buffer. Bossi et al. (1994a,b) produced IPGs which spanned the pH 10–12 interval by incorporating a very basic acrylamido buffer with a pK value > 13, and rehydrating the IPG gel in a 0–10% sorbitol gradient to limit the electroendosmotic flow (EOF). At very high pH the polyacrylamide matrix is positively charged and a strong EOF to the anode is created, which is opposite to the direction of the EOF typically observed when the gel matrix is negatively charged at lower pH. By rehydrating the IPG in a 0–10% (anode to cathode) gradient of sorbitol, the EOF can be partially quenched. However, when pH 10–12 IPG gels were used for separating complex mixtures of basic ribosomal proteins from rat liver and HeLa cells, severe horizontal streaking caused by the reverse EOF limited the resolution (Görg et al. 1997). Replacing the sorbitol gradient with 16% isopropanol (or 10% isopropanol, 0.2% methylcellulose) in the IPG rehydration solution quenched the reverse EOF and allowed steady state focusing with most proteins appearing as well resolved round spots (Görg et al. 1997). In addition to the use of isopropanol and methylcellulose, the IPG gels were prepared using dimethyl acrylamide instead of acrylamide to increase the stability of the matrix at high pH.

A further difficulty associated with the focusing of basic proteins in IPG gels is the loss of thiol reducing agents through migration during the IEF, which can require up to 35 kVh. As discussed in Sect. 2.2, TBP is non-charged and thus does not migrate during the IEF and as such may provide a further enhancement to the focusing of basic proteins by increasing the solubility and speeding the focusing process. The incorporation of thiourea may also prove beneficial in the separation of basic proteins to ensure a minimum of hydrophobic interactions between the sample proteins and the IPG.

2.6 Protein quantitation

Most physiological and pathological processes are associated with quantitative variations in the amounts of gene products. Studies which are undertaken at the DNA or mRNA levels do not provide a complete picture because there is not a direct correlation between the levels of mRNA and the levels of protein in cells (Anderson and Seilhamer 1997). Since the discovery of 2-D PAGE, many studies of protein quantitation have been undertaken (Anderson et al. 1985; Anderson et al. 1987; Bossinger et al. 1979; Garrels 1983; Giometti et al. 1987; Giometti et al. 1990). The most widely used technique to detect protein spots on two-dimensional gels is silver staining. It is well known that this procedure is far from stoichiometric, but the intensity is linear over 40–50 fold range in concentration from 0.04 ng/mm^2 to 2 ng/mm^2. This value compares well with the 20-fold range (10–200 ng) of linearity found for Coomassie Blue staining (Dunn 1987). Above this limit the stain density becomes non-linear as spot densities reach saturation (Merril et al. 1984; Righetti 1990). In a study on human leukocyte proteins, Giometti et al. (1991) demonstrated that 200 silver stained spots were observed to have a coefficient of variation (standard deviation divided by the mean × 100) above 15%, but that in dilution experiments the majority of detected proteins have a linear relationship between the amount of protein loaded and the spot volume.

Another issue which limits the use of silver staining for overall quantitation is that the relationship between silver staining density and protein concentration is characteristic for each protein (Merril et al. 1984; Giometti et al. 1991). The amino acid composition and post-translational modifications of a protein spot determine the final quantity of silver which is observed in the gel. It must also be noted that protein-specific staining curves have been also observed for organic stains, including Coomassie Blue (Tal et al. 1985). Merril et al. (1993) suggested the use of constitutively expressed proteins as internal standards for the normalisation of spot intensity to overcome this problem in comparative studies.

Some recent studies on acute phase proteins during bacterial and viral diseases (Bini et al. 1996) or in perinatal human plasma (Liberatori et al. 1997) discuss the use of quantitation to clarify specific pathological problems. Bini et al. (1996) used a laser densitometer and the Melanie II[3] software for 2-D gel matching to perform a comparison of pathological with recovery sera and match homologous protein spots in both gels. To overcome the problem of stain variability from gel to gel, measures such as relative volume ratios (spot volume divided by total spot volumes) or relative optical density ratios between 2 sets of 6 gels were used (Bini et al. 1996).

Other detection procedures, such as radiolabelling with specific isotopes, have been applied to detect and quantify polypeptides separated by 2-D electrophoresis. Some groups have developed procedures to calibrate films used for autoradiography or fluorography (Bossinger et al. 1979; Garrels 1983). The more recent introduction of phosphor-imaging technology has further enhanced the sensitivity and linearity of radiolabel detection methods. Radiolabelling can be a highly sensitive detection method, but it has limitations because the amount of label incorporated by each protein is dependent on the protein amino acid composition. An even more serious drawback is that radiolabelling is limited to samples which are able to be labelled during culture, which may require specialised facilities to handle large quantities of isotopes. Many countries have adopted stringent laws which restrict the use of radioisotopes and make this type of approach less favourable.

Considering the above difficulties with protein staining, protein quantitation remains a challenge which must be solved such that 2-D PAGE databases can be extended to consider protein abundances as components of pathological responses. An alternative approach to quantitation is to use staining only as a means to localise spots after 2-D PAGE and blotting to PVDF. Wilkins et al. (1996) proposed the use of high throughput rapid amino acid analysis for protein quantitation on PVDF blots. Amino acid analysis measures the protein composition and thus can determine the quantity of a particular protein in molar or gram terms.

2.7 Future directions for 2-D PAGE

In the last twenty-five years many technical advances have been implemented to make 2-D PAGE a very powerful method for the separation and purification of complex protein mixtures. However, for 2-D PAGE to fulfil its potential the technology must be simplified to allow novice and casual users to achieve high quality separations. Currently one dimensional SDS-PAGE is the standard method used in most laboratories to assess protein purity and often to separate proteins for analyses such as Edman sequencing, after blotting to PVDF membranes (Matsudaira 1987). We suggest that the current goal for 2-D PAGE developments should be to achieve a technical simplicity that will allow 2-D to take over from one dimensional SDS-

---

[3] http://www.expasy.ch/melanie/melanie-top.html

PAGE as the standard laboratory electrophoresis method. This simplification presents a number of challenges, to which some potential solutions are presented below.

2.7.1 Simplifying the IPG-SDS-PAGE interface

In the preceding sections we have discussed the benefits of sample application during sample rehydration and the use of highly solubilising reagents such as thiourea and TBP. These advances in 2-D PAGE methodology have broadened the scope of the technique by providing near universal methods for sample solubilisation and loading. Before this, sample preparation for 2-D PAGE was a highly individual procedure, with each sample requiring special treatment. In contrast to the excellent progress with solubilisation and loading, there are other steps, such as the IPG-SDS-PAGE interface, which have remained cumbersome and labour intensive. As discussed in Sect. 2.2.1, one of the drawbacks of IPG gels is the tendency of some proteins to adsorb to the matrix, thus resulting in proteins which are absent or under-represented on the second dimension gel. When using the conventional 2-D PAGE methodology, it is necessary to equilibrate the IPG gel in a strongly denaturing solution after IEF to re-solubilise proteins and ensure adequate transfer of proteins to the second dimension gel (Görg et al. 1987). The typical equilibration involves soaking the IPG in 6 M urea, 2% SDS and 2% DTT for 10 minutes and then a further 10 minute soaking in a similar solution where DTT is replaced with 2.5% iodoacetamide. The iodoacetamide alkylates any free DTT, thus minimising vertical streaking in the second dimension that is caused by DTT migrating through the SDS-PAGE gel (Görg et al. 1987). This process is cumbersome and the exact timing of the two steps should ideally be empirically determined for each different sample. If the equilibration is too long proteins can be lost from the IPG gel and resolution decreased. Conversely, an equilibration that is too short will result in poor protein transfer to the SDS-PAGE gel. The solution is to eliminate the equilibration steps completely and replace them with a simplified protocol.

In vertical SDS-PAGE the IPG is embedded in molten agarose on the top of the second dimension gel. By incorporating SDS and a reducing agent into the agarose it is possible to solubilise the protein in the IPG and ensure an efficient transfer to the second gel. However, DTT in the second dimension gel can also cause vertical streaking by solubilising dust trapped in the gel during casting. The equilibration during agarose embedding can be enhanced by the use of TBP in the IPG and in the agarose, because TBP is not charged and thus does not migrate in the SDS gel to cause vertical streaking during the second dimension gel run.

2.7.2 Fluorescent protein detection

Fluorescent stains such as SYPRO Orange and Red (Steinberg et al. 1996) have a number of advantages over conventional staining techniques such as silver and Coomassie Blue. Visualising the proteins after separation is a relatively short procedure which can usually be completed within 30 minutes, in contrast to silver or Coomassie Blue staining which take from 2 hours to overnight to complete. During SYPRO Orange and Red staining the proteins are not fixed in the gel, so blotting to membranes can be done after staining. However, correct staining requires that the proteins are coated in SDS, so the drawbacks that affect silver staining are also inherent to this type of stain. SDS binding to proteins is affected by amino acid composition and in particular, post-translational modifications such as glycosylation. Therefore SYPRO dye binding is not uniform for all the protein species on a 2-D PAGE gel.

Labelling proteins with fluorescent reagents prior to the IEF has also been used for protein detection, but most reagents introduce charges to proteins and therefore cause shifts in pI (Righetti 1990). This is a serious drawback when one wishes to compare a 2-D gel with a reference map in a database such as SWISS-2DPAGE (see Sect. 5.5.1 and Chap. 6). The ideal time for labelling proteins is during the transfer between the first and second dimension, as the proteins have focused to their pIs and the mass change due to the fluorescent label is not very significant in the second dimension. During the transfer between the IPG and the SDS gel the proteins are in concentrated zones in the IPG and then undergo stacking in the initial stage of the SDS-PAGE run, which means that minimal amounts of labelling reagents are required. Jackson et al. (1988) used an ε-amino group labelling reagent, 2-methoxy-2,4-diphenyl-3(2H)furanone, to form fluorescent protein derivatives after the first dimension IEF. Proteins were visualised after the SDS-PAGE using an ultraviolet transilluminator and CCD camera. Previously, fluorescent alkylation has been difficult to achieve because of the presence of high concentrations of DTT in the conventional equilibration protocol. Omitting DTT and substituting TBP in the IEF, and in the transfer as discussed above, it is possible to alkylate proteins with fluorescent reagents such as fluorescein maleimide (Herbert et al. 1997).

A major advantage of fluorescent labelling during the transfer from first to second dimension gels is that no fixing or further staining is required to visualise the separated proteins. Gels can be imaged and then blotted to membranes if desired, or in-situ enzymatic digestions can be performed with the prospect of higher peptide yields than from proteins fixed in gels.

2.7.3 High throughput 2-D PAGE

To take advantage of the resolving power of 2-D PAGE for screening clinical samples, rapid separations and high throughput are necessary. Although multi-gel SDS-

PAGE systems for up to 20 gels are available, these use large format gels (i.e. 25 × 25 cm) and they are very expensive and slow to use for screening purposes. Currently large format 2-D PAGE gels take between 3 and 5 days to complete and are quite labour intensive.

To simplify 2-D PAGE and increase reproducibility it is preferable to use pre-made SDS-PAGE gels and premade IPGs, the latter of which are already commercially available in large format. This shifts the onus for quality control to a professional instrument company and should improve the inter-laboratory reproducibility. Increased protein solubility using TBP and thiourea decreases the IEF run time and gives improved separations. Other technical innovations such as sample application during IPG rehydration and equilibration during agarose embedding have simplified the 2-D PAGE process and decreased the overall time required. By combining these elements in a mini 2-D PAGE system it is now possible to complete a 2-D gel run in one working day. Fig. 2.4 (see p. 133) shows a mini 2-D PAGE gel that was completed from IPG rehydration to a finished Coomassie Blue stain in less than 9 hours. The ability to complete 2-D PAGE gels in one day opens the way for high throughput 2-D for screening samples, optimising sample preparation protocols or preparative purification of prefractionated proteins.

2.8 Conclusion

In the time since Kenrick and Margolis (1970) and others began pioneering 2-D PAGE separations, many technical advances have been made which now make 2-D PAGE a vital tool in proteomics and modern biology. Whilst 2-D separations first involved native proteins, modern 2-D separations almost exclusively use reducing conditions. This is appropriate for proteome definition, which involves making maps of all proteins expressed. We can foresee that the technology may soon turn a full circle and native 2-D PAGE will assume a complementary role as the study of higher order protein-protein interactions becomes more and more important.

Acknowledgments

We are grateful to Mark Molloy and Mark Broughton for supplying the 2-D gels for Fig. 2.1 and 2.4. BH acknowledges the support of the Australian Proteome Analysis Facility, Macquarie University Centre for Analytical Biotechnology, the Wool Research Organisation of New Zealand, the Australian Medical Research Council, and the Australia Research Council.

References

Anderson L, Seilhamer J (1997) A comparison of selected mRNA and protein abundances in human liver. Electrophoresis 18:533–537

Anderson NL, Nance SL, Tollaksen SL, Giere FA, Anderson NG (1985) Quantitative reproducibility of measurements from Coomassie Blue-stained two-dimensional gels: analysis of mouse liver protein patterns and a comparison of BALB/c and C57 strains. Electrophoresis 6:592–599

Anderson NL, Giere FA, Nance SL, Gemmell MA, Tollaksen SL, Anderson NG (1987) Effects of toxic agents at the protein level: quantitative measurement of 213 mouse liver proteins following xenobiotic treatment. Fundam Appl Toxicol 8:39–50

Bier M, Egen NB, Allgyer TT, Twitty GE, Mosher RA (1979) In: Gross E, Meienhofer J (eds) Peptides: structure and biological functions. Pierce Chemical Co Rockford, Illinois, pp 79–89

Bini L, Magi B, Marzocchi B, Cellesi C, Berti B, Raggiaschi R, Rossolini A, Pallini V (1996) Two-dimensional electrophoretic patterns of acute-phase human serum proteins in the course of bacterial and viral diseases. Electrophoresis 17:612–616

Bjellqvist B, Ek K, Righetti PG, Gianazza E, Görg A, Westermeier R, Postel W (1982) Isoelectric focusing in immobilised pH gradients: principle, methodology and some applications. J Biochem Biophys Meth 6:317–339

Bjellqvist B, Pasquali C, Ravier F, Sanchez JC, Hochstrasser D (1993a) A nonlinear wide-range immobilised pH gradient for two-dimensional electrophoresis and its definition in a relevant pH scale. Electrophoresis 14:1357–1365

Bjellqvist B, Sanchez JC, Pasquali C, Ravier F, Paquet N, Frutiger S, Hughes GJ, Hochstrasser D (1993b) Micropreparative 2-D electrophoresis allowing the separation of milligram amounts of proteins. Electrophoresis 14:1375–1378

Blomberg A, Blomberg L, Norbeck J, Fey SJ, Larsen PM, Larsen M, Roepstorff P, Degand H, Boutry M, Posch A, Görg A (1995) Interlaboratory reproducibility of yeast protein patterns analysed by immobilised pH gradient two-dimensional gel electrophoresis. Electrophoresis 16:1935–1945

Bossi A, Gelfi C, Orsi A, Righetti PG (1994a) Isoelectric focusing of histones in extremely alkaline immobilised pH gradients: comparison with capillary electrophoresis. J Chromatogr A 686:121–128

Bossi A, Righetti PG, Vecchio G, Severinsen S (1994b) Focusing of alkaline proteases (subtilisins) in pH 10–12 immobilised gradients. Electrophoresis 15:1535–1540

Bossinger J, Miller MJ, Vo KP, Geiduschek EP, Xuong NH (1979) Quantitative analysis of two-dimensional electrophoretograms. J Biol Chem 254:7986–7998

Celis JE, Rasmussen HH, Madsen P, Leffers H, Honoré B, Dejgaard K, Gesser B, Olsen E, Gromov P, Hoffmann HJ, Nielsen M, Celis A, Basse B, Lauridsen JB, Ratz GP, Nielsen H, Andersen AH, Walbrum E, Kjaergaard I, Puype M, Van Damme J, Vandekerckhove J (1992) The human keratinocyte two-dimensional gel protein database (update 1992): towards an integrated approach to the study of cell proliferation, differentiation and skin diseases. Electrophoresis 13:893–959

Celis, JE, Rasmussen HH, Gromov P, Olsen E, Madsen P, Leffers H, Honoré B, Dejgaard K, Vorum H, Kristensen DB, Ostergaard M, Haunso A, Jensen NA, Celis A, Basse B, Lauridsen J, Ratz GP, Andersen AH, Walbum E, Kjaergaard I, Andersen I, Puype M, Van

Damme J, Vandekerckhove J (1995) The human keratinocyte two-dimensional gel protein database (update 1995): mapping components of signal transduction pathways. Electrophoresis 14:1091–1098

Corbett JM, Dunn MJ, Posch A, Görg A (1994) Positional reproducibility of protein spots in two-dimensional polyacrylamide electrophoresis using immobilised pH gradient isoelectric focusing in the first dimension — an interlaboratory comparison. Electrophoresis 15:1205–1211

Corthals GL, Molloy MP, Herbert BR, Williams KL, Gooley AA (1997) Prefractionation of protein samples prior to two-dimensional electrophoresis. Electrophoresis 18:317–323

Dunn MJ (1987) Two-dimensional polyacrylamide gel electrophoresis. In: Chrambach A, Dunn MJ, Radola BJ (eds) Advances in electrophoresis. VCH, Weinheim, Germany, pp 1–109

Dunn MJ (1993) Gel electrophoresis: proteins. Bios Scientific Publishers, Oxford, UK

Eckerskorn C, Strupat K, Schleuder D, Hochstrasser DF, Sanchez JC, Lottspeich F, Hillenkamp F (1997) Analysis of proteins by direct scanning-infrared-MALDI mass spectrometry after 2-D PAGE separation and electroblotting. Anal Chem in press

Esteve-Romero J, Simo-Alfonso E, Bossi A, Bresciani F, Righetti PG (1996) Sample streaks and smears in immobilised pH gradient gels. Electrophoresis 17:704–708

Garrels JI (1983) Quantitative two-dimensional gel electrophoresis of proteins. Methods Enzymol 100:411–423

Giometti CS, Gemmell MA, Nance SL, Tollaksen SL, Taylor J (1987) Detection of heritable mutations as quantitative changes in protein expression. J Biol Chem 262:12764–12767

Giometti CS, Tollaksen SL, Gemmell MA, Taylor J, Hawes J, Roderick T (1990) The analysis of recessive lethal mutations in mice by using two-dimensional gel electrophoresis of liver proteins. Mut Res 242:47–55

Giometti CS, Gemmell MA, Tollaksen SL, Taylor J (1991) Quantitation of human leukocyte proteins after silver staining: a study with two-dimensional electrophoresis. Electrophoresis 12:536–543

Görg A, Postel W, Weser J, Gunther S, Strahler JR, Hanash SM, Sommerlot L (1987) Elimination of point streaking on silver stained two-dimensional gels by addition of iodoacetamide to the equilibration buffer. Electrophoresis 8:122–124

Görg A, Postel W, Gunther S (1988) The current state of two-dimensional electrophoresis with immobilized pH gradients. Electrophoresis 9:531–546

Görg A, Obermaier C, Boguth G, Csordas A, Diaz JJ, Madjar JJ (1997) Very alkaline immobilised pH gradients for two-dimensional electrophoresis of ribosomal and nuclear proteins. Electrophoresis 18:328–337

Hanash SM, Strahler JR, Neel JV, Hailat N, Melhem R, Keim D, Zhu XX, Wagner D, Gage DA, Watson JT (1991) Highly resolving two-dimensional gels for protein sequencing. Proc Natl Acad Sci USA 88:5709–5713

Herbert BR, Molloy MP, Gooley AA, Walsh BJ, Bryson WG, Williams KL (1997) Improved protein solubility in 2-D electrophoresis using tributyl phosphine. Submitted

Hochstrasser DF, Merril CR (1988) "Catalysts" for polyacrylamide gel polymerization and detection of proteins by silver staining. Appl Theor Electrophor 1:35–40

Huber LA, Pasquali C, Gagescu R, Zuk A, Gruenberg J, Matlin KS (1996) Endosomal fractions from viral K-ras-transformed MDCK cells reveal transformation specific changes on two-dimensional gel maps. Electrophoresis 17:1734–1740

Jackson P, Urwin VE, Mackay CD (1988) Rapid imaging, using a cooled charged-coupled-device, of fluorescent two-dimensional polyacrylamide gels produced by labelling proteins in the first dimensional isoelectric focusing gel with the fluorophore 2-methoxy-2,4-diphenyl-3(2H)furanone. Electrophoresis 9:330–339

Kenrick KG, Margolis J (1970) Isoelectric focusing and gradient gel electrophoresis: a two-dimensional technique. Anal Biochem 33:204–207

Klose J (1975) Protein mapping by combined isoelectric focusing and electrophoresis in mouse tissues. A novel approach to testing for induced point mutations in mammals. Humangenetik 26:231–243

Liberatori S, Bini L, De Felice C, Magi B, Marzocchi B, Raggiaschi R, Pallini V, Bracci R (1997) Acute-phase proteins in perinatal human plasma. Electrophoresis 18:520–526

Matsudaira P (1987) Sequence from picomole quantities of proteins electroblotted onto poly-vinylidene difluoride membranes. J Biol Chem 262:10035–10038

Merril CR, Goldman D, Van Keuren ML (1984) Gel protein stains: silver stain. Methods Enzymol 104:441–447

Merril CR, Creed GJ, Joy J, Olson AD (1993) Identification and use of constitutive proteins for the normalization of high resolution electrophoretograms. Appl Theor Electrophor 3:329–333

O'Farrell PH (1975) High resolution two-dimensional electrophoresis of proteins. J Biol Chem 250:4007–4021

Rabilloud T (1994) Two-dimensional electrophoresis of basic proteins with equilibrium isoelectric focusing in carrier ampholyte pH gradients. Electrophoresis 15:278–282

Rabilloud T (1996) Solubilisation of proteins for electrophoretic analyses. Electrophoresis 17:813–829

Rabilloud T, Hubert M, Tarroux P (1986) Procedures for two-dimensional electrophoretic analysis of nuclear proteins. J Chromatogr 351:77–89

Rabilloud T, Valette C, Lawrence JJ (1994) Sample application by in-gel rehydration improves the resolution of two-dimensional electrophoresis with immobilised pH gradients in the first dimension. Electrophoresis 15:1552–1558

Rabilloud T, Adessi C, Giraudel A, Lunardi J (1997) Improvement of the solubilisation of proteins in two-dimensional electrophoresis with immobilised pH gradients. Electrophoresis 18:307–316

Righetti PG (1990) Immobilised pH gradients: theory and methodology. In: Burdon RH, van Knippenberg PH (eds) Laboratory techniques in biochemistry and molecular biology. Elsevier, Amsterdam

Righetti PG, Wenisch E, Faupel M (1989) Preparative protein purification in a multi-compartment electrolyser with Immobiline membranes. J Chromatogr 475:293–309

Sanchez JC, Paquet N, Hughes GJ, Hochstrasser DF (1992) Preparative 2-D purifies proteins for sequencing or antibody production. US/EG Bio-Rad Bulletin 1744

Sanchez JC, Rouge V, Pisteur M, Ravier F, Tonella L, Moosmayer M, Wilkins MR, Hochstrasser DF (1997a) Improved and simplified in-gel sample application using reswelling of dry immobilised pH gradients. Electrophoresis 18:324–327

Sanchez JC, Wirth P, Jaccoud S, Appel RD, Sarto C, Wilkins MR, Hochstrasser DF (1997b) Simultaneous analysis of cyclin and oncogene expression using multiple monoclonal antibody immunoblots. Electrophoresis 18:638–641

Scheele GA (1975) Two-dimensional gel analysis of soluble proteins. Characterisation of guinea pig exocrine pancreatic proteins. J Biol Chem 250:5375–5385

Steinberg TH, Haugland RP, Singer VL (1996) Applications of SYPRO Orange and SYPRO Red protein gel stains. Anal Biochem 239:238–245

Tal M, Silberstein A, Nusser E (1985), Why does Coomassie Brilliant Blue R interact differently with different proteins? J Biol Chem 260:9976–9980

Wilkins MR, Sanchez JC, Williams KL, Hochstrasser DF (1996) Current challenges and future applications for protein maps and post-translational vector maps in proteome projects. Electrophoresis 17:830–838

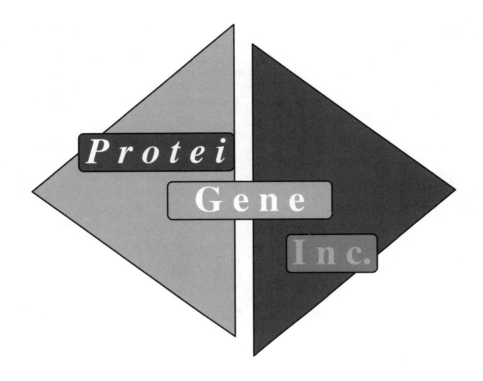

Leaders in Proteomics
by Mass Spectrometry

Fast, sensitive, reliable analysis with ProteiGene's proprietary
PCS ™ -- the Proteome Characterization System.

ProteiGene, Inc.
44 Manning Road
Billerica, MA 01821
USA

How to reach us:
Voice 508-670-1761
Fax   508-670-1762
bioservices@proteigene.com

3 Protein Identification in Proteome Projects

Marc R. Wilkins and Andrew A. Gooley

3.1 The purpose of protein identification

Two-dimensional gel electrophoresis in its modern form was described in 1975 (O'Farrell 1975; Klose 1975). Even the earliest papers showed that thousands of proteins could be resolved on a single gel, and that qualitative differences between samples could be detected. Researchers excitedly commented that such gels potentially represented a means of investigating global processes in living systems. However, in many ways the invention of 2-D separation was ahead of its time. For example, molecular biology was in its infancy. The cloning of cDNA had just been described (Rougeon et al. 1975), but methods like Sanger DNA sequencing (Sanger et al. 1977) or the polymerase chain reaction (Saiki et al. 1985) were unknown. By contrast, protein sequencing by Edman degradation had been defined and automated (Edman and Begg 1967) but still required hundreds of picomoles of protein — at least an order of magnitude more than that purified on a 2-D gel. The consequence of the above factors was that protein and DNA sequence was only available for a handful of proteins. The first genome sequence was still 20 years away! So whilst in 1975 2-D gels could separate a sample into thousands of components, and could perhaps describe interesting phenomena in terms of spots on a gel, without protein identification these gels remained little more than a tool of description or classification.

More recently, things have changed. New genomes are completed almost monthly, analytical techniques are exquisitely accurate and sensitive, and computers have made the management of massive amounts of sequence and other biochemical data feasible. Together, this means that the identification of proteins on 2-D gels is increasingly a routine undertaking. But in the current era of biology, what does the identification of a protein, purified by 2-D gel electrophoresis or otherwise, actually mean?

In proteome projects, which aim to identify and characterise all proteins expressed by an organism or tissue (Wilkins et al. 1995), the identification of proteins is central. In the simplest sense, identification assigns a name or database accession code to a protein or spot on a gel, thus linking amino acid sequences of proteins to DNA sequences of genes, and linking genomes to proteomes. Identification is also the first step towards studies on protein co- and post-translational modification (see Chap. 4), and ultimately, function. For proteins on 2-D gels, identifications create "reference 2-D maps" which begin to define exactly what proteins are expressed in an organism or tissue. Reference maps of different tissues of an organism, in turn, can reveal proteins unique to a tissue, and define which ones are common. When combined with 2-D gel studies on protein complexes (e.g. Neubauer et al. 1997), identifications allow the understanding of which proteins are covalently or non-covalently associated, thus providing a powerful means for examining how the many proteins of a cell interact in a coordinated manner.

On a molecular level, the identification of proteins on 2-D gels provides a means to verify information generated in genome projects. The confirmation of open reading frame predictions, including those for "tiny" proteins of less than 100 amino acids, is a primary objective, as is the checking for frameshift mutations (see Chap. 5). Protein post-translational processing, some of which can be predicted from translated DNA sequences, can be checked through protein identification if protein N- and C-termini are studied, and mRNA splicing variation may be investigated through detailed protein primary structure analysis. Finally, identification procedures can supply quantitative data for levels of proteins on gels, which should ultimately be useful in conjunction with data from mRNA microarray analysis (see Schena 1996) to better understand not only where and when a protein is likely to be expressed, but also parameters such as protein copy number per cell, pool size and half life. These data bring us closer to the definition of what the component molecules of a cell are, and thus the components necessary for a living unit.

3.2 An overview of protein identification strategies

Given that the best way of currently separating and visualising a proteome is through 2-D gel electrophoresis, the challenge that then arises is how the hundreds to thousands of proteins of a proteome can be identified with a minimum of effort and in a cost- and time-effective manner. Here we will not consider the more traditional protein identification methods such as immunoblotting (see Sect. 8.7.6), chemical sequencing of internal peptides, comigration analysis of known and unknown proteins, or overexpression of homologous genes of interest in an organism under study. This is not because they are ineffective identification techniques, but only because they are generally slow, labour intensive and/or expensive, and thus unsuitable for use as high throughput approaches. Alternatively, we will consider the range of different protein attributes that can be defined analytically, and how these can be used by themselves, or in combinations, to identify any particular protein.

The identification of proteins involves the definition of one or more attributes of proteins (Table 3.1), which are then matched against protein databases in various manners. We think of protein attributes as primary or secondary: primary if it is a property of or generated directly from the intact protein; secondary if it represents or is generated from fragments of the whole molecule. Most attributes relate directly or indirectly to a protein's sequence, however they vary in the way that they are generated and the protein property they represent. The attributes also vary widely in their usefulness as unique identifiers and in how quickly they can be determined, which affects analytical throughput. In the following sections of this chapter the use of protein attributes for identification in proteome projects will be explored, as well as some technical aspects of these identification procedures. For detailed technical notes on identification methods, the reader is referred elsewhere to a comprehensive review by Aebersold and Patterson (1997), and the recent book "2-D Protein Gel Electrophoresis Protocols" (1997), edited by A.J. Link.

Table 3.1. Primary and secondary protein attributes that can be used for identification by matching against databases. References for techniques are in appropriate sections of the text

Protein attribute	Source of analytical data	Analysis type
Primary:		
species of origin	—	—
isoelectric point	gel image analysis	parallel[a]
apparent mass	gel image analysis	parallel
mass	direct MALDI-TOF MS[b] of protein immobilised on membrane or in gel matrix	serial
protein N- and C-terminal sequence tag	chemical sequencing of protein immobilised on membrane	parallel or serial
extensive N-terminal protein sequence	chemical sequencing of protein immobilised on membrane	serial
Secondary:		
peptide mass fingerprinting	MALDI or ESI-MS[c] analysis of peptides following digestion of proteins from gels or immobilised on membranes	serial
peptide fragmentation data and *de novo* MS sequencing	fragmentation of peptides from mass fingerprinting procedures	serial
amino acid composition	multiple radiolabelling analysis; chromatographic analysis of hydrolysed proteins	parallel; serial

[a]meaning that this attribute can be determined for many proteins at the same time
[b]Matrix-Assisted Laser Desorption / Ionisation - Mass Spectrometry
[c]Electrospray Ionisation - Mass Spectrometry

3.3 Primary attributes for protein identification

3.3.1 Protein species of origin

A fundamental property of any protein is its species of origin. However, this is frequently overlooked because until recently most identification strategies matched
data against all species in databases in order to maximise the chance of finding a
potential identity for a protein. With the availability of complete proteomes in databases, including *Mycoplasma genitalium*, *Escherichia coli*, *Haemophilus influenzae*, *Methanococcus jannaschii*, *Streptococcus pneumoniae*, *Staphylococcus
aureus*, *Archeobacter globus*, *Treponema pallidium*, *Helicobacter pylori*, and *Saccharomyces cerevisiae*, it is now feasible and often desirable to integrate species of
origin into identification procedures (Sect. 5.2.1). This can be done by single-spe-
cies matching, where, for example, data for proteins from *E. coli* are matched only
against *E. coli* proteins in the database, or by cross-species matching, where proteins from, for example, *Candida albicans*, are matched only against the proteome
of another organism in the kingdom Fungi such as *S. cerevisiae*. In these cases, the
net effect is to match against only 5,000 or so proteins, thus eliminating the "noise"
from many tens of thousands of other proteins from unrelated species. Many
searching tools are now capable of doing single- or cross-species matching, and
some (e.g. those from ExPASy for searching the SWISS-PROT protein database)
can match against all proteins from a chosen genus, family, phylum or kingdom.
Cross-species matching is further explored in Sect. 3.5.

3.3.2 Protein isoelectric point

The isoelectric point (pI) of a denatured protein is a parameter determined by its
amino acid composition, N- and C-terminal amino acids, and any post-translational
modifications. On 2-D gels, the pI of all proteins can be estimated at the same time,
making pI an attribute obtainable in a parallel, rather than serial fashion. Estimation is done by first calculating the theoretical pI of 10 or more known "landmark"
proteins on a 2-D gel, which can be done using the ExPASy Compute pI/MW tool
(Table 3.2), which applies pK values that Bjellqvist et al. (1993, 1994) defined for
proteins run on immobilised pH gradients. Theoretical pI values for known spots
can then be used within gel image analysis systems to create a pI "grid", allowing
the pI of other proteins on the gel to then be assigned. The accuracy of these estimates will depend largely on the care that is taken in the construction of the grid,
and the type of sample being studied. Only proteins known to be unmodified
should be used as landmarks, and large proteins should be used in preference to
those that are small to avoid inaccuracies associated with pI predictions for proteins of low buffer capacity. In this manner, pI estimates for denatured unmodified cytoplasmic proteins separated on 2-D gels can usually be estimated to ± 0.25

Table 3.2. Protein identification and characterisation programs available on the Internet, and their URL addresses. Identification programs are grouped according to the protein attributes they use to query databases. An up-to-date list is at http://www.expasy.ch/tools.html

Type of Program / Name	Internet URL address
Intact protein molecular mass:	
PeptideSearch	http://www.mann.embl-heidelberg.de/Services/ PeptideSearch/PeptideSearchIntro.html
TagIdent	http://www.expasy.ch/www/guess-prot.html
Protein sequence tags:	
PeptideSearch	http://www.mann.embl-heidelberg.de/Services/ PeptideSearch/FR_SequenceOnlyForm.html
TagIdent	http://www.expasy.ch/www/guess-prot.html
Peptide mass fingerprinting:	
MassSearch	http://cbrg.inf.ethz.ch/subsection3_1_3.html
MS-Fit	http://falcon.ludwig.ucl.ac.uk/MS-Fit.html
PeptideSearch	http://www.mann.embl-heidelberg.de/Services/ PeptideSearch/FR_PeptideSearchForm.html
ProFound	http://chait-sgi.rockefeller.edu/cgi-bin/prot-id
Peptide mass fingerprinting and sequence tag data:	
MS-Edman	http://falcon.ludwig.ucl.ac.uk/msedman.htm
Mowse	http://gserv1.dl.ac.uk/SEQNET/mowse.html
Peptide sequence tags (from MS/MS or MALDI-PSD):	
MS-Tag	http://falcon.ludwig.ucl.ac.uk/mstag.htm
PepFrag	http://chait-sgi.rockefeller.edu/cgi-bin/prot-id-frag
PeptideSearch	http://www.mann.embl-heidelberg.de/Services/ PeptideSearch/FR_PeptidePatternForm.html
Sequest[a]	http://thompson.mbt.washington.edu/sequest.html
Amino acid composition (with or without sequence tag):	
AACompIdent	http://www.expasy.ch/ch2d/aacompi.html
PropSearch	http://www.embl-heidelberg.de/aaa.html
Amino acid composition, protein sequence tag, peptide mass fingerprinting:	
MultiIdent	http://www.expasy.ch/sprot/multiident.html
Programs that assist interpretation of analytical data:	
Amino Acid Sequence	http://chait-sgi.rockefeller.edu/cgi-bin/sequence
Compute pI/MW	http://www.expasy.ch/ch2d/pi_tool.html
MS-Digest	http://falcon.ludwig.ucl.ac.uk/msdigest.htm
PeptideMass	http://www.expasy.ch/sprot/peptide-mass.html

[a]This program cannot be queried via the WWW to obtain results. Users must download the program and install it locally.

units of that predicted (Bjellqvist et al. 1993, 1994; Wilkins et al. 1996a). Clearly it is not possible to identify a protein solely on the basis of estimated pI, however judicious use of this parameter can exclude many proteins in databases during identification procedures. Therefore estimated protein pI is best used in combination with other protein attributes to achieve identification.

3.3.3 Protein apparent mass and mass

The mass of a whole protein is one of the most commonly used attributes in protein identification procedures. Protein mass is determined by the total mass of all amino acids which constitute the protein, plus the mass of any post-translational modifications. There are two ways in which the mass of proteins from 2-D gels can be determined. Firstly this can be done by gel image analysis (see Sect. 6.4), whereby the theoretical mass of known protein spots are calculated from the database, and these proteins then used within a gel image analysis program to create a grid of apparent protein mass on the gel. As with estimated protein pI, this procedure is parallel, supplying values of apparent mass for up to thousands of proteins simultaneously. However, many proteins migrate anomalously under conditions of SDS-PAGE, and apparent mass values for unmodified, small proteins may be in error by more than 30% (e.g. Wilkins et al. 1996a). Mass estimations for post-translationally modified proteins from gels are even more problematic, with many glycoproteins having a Stokes radius that gives an apparent mass 50% larger than that predicted from the database (e.g. Gooley et al. 1997).

The second and most accurate technique for determining the mass of proteins from gels is by mass spectrometry. In this approach, proteins are electroblotted from gels to a membrane (e.g. nitrocellulose or PVDF), following which the masses of intact proteins are measured. This may involve dissolving pieces of nitrocellulose to liberate proteins, whose masses can then be measured by standard matrix-assisted laser desorption / ionisation - mass spectrometry (MALDI-TOF MS) techniques (Liang et al. 1996). More elegantly, PVDF membranes carrying many proteins may be overlaid with matrix, and placed directly into a MALDI mass spectrometer, which is generally equipped with an infrared laser (Eckerskorn et al. 1997; Sutton et al. 1997; Vestling and Fenselau 1994). In its most advanced form, this allows membranes to be scanned at high resolution, thus in principle generating mass data for many proteins and constructing pseudo-images of the gel separation (Eckerskorn et al. 1997). However, it remains unclear if proteins larger than 100 kDa can be analysed with this approach.

Considering that the accuracy obtainable with mass spectrometry for unmodified proteins is usually within 1% of that predicted from the database, can proteins be identified by their mass alone, or in combination with their estimated pI? This will depend upon the proteome size of the organism being studied, and the size and charge of the protein. In *E. coli*, which has 4,285 proteins predicted from its genomic sequence, a contour map (Fig. 3.1, see p. 134) can demonstrate the theoretical number of proteins found in a certain pI and mass region of a 2-D gel. From this map we would predict that a protein 2-D spot with pI 5.5 and mass 25,000 is unlikely to be identified by these attributes because there are tens of other proteins known to be within this region. However, a protein with pI 7.2 and mass 70,000 perhaps could be uniquely identified because there are very few proteins in *E. coli* of this size and charge. Yet having said that, there are few who would be convinced that a protein can be confidently identified by the use of pI and mass values alone. These parameters are better used in combination with other protein attributes, or to generate lists of proteins from databases that may include the correct protein identification. Two tools, PeptideSearch and TagIdent, are available on the Internet to generate such lists (Table 3.2). A sample output from one of these shows that pI and MS-determined mass values for a human heart protein (Sutton et al. 1997) can generate a short list of proteins that contains the correct identification, but cannot give an indication as to what the protein identity actually is (Fig. 3.2).

3.3.4 Protein N- and C-terminal sequence tags

Protein sequence, even if only a few amino acids in length, is very specific. For example, there are 8,000 possible combinations of sequences of 3 amino acids, 160,000 combinations of sequences of 4 amino acids, and a massive 3,200,000 combinations of sequences 5 amino acids long. Small stretches of sequence were first used for database searching by Mann and Wilm (1994), who found they could be generated by tandem mass spectrometry fragmentation of peptides. They termed these "peptide sequence tags" (see Sect. 3.4.1 for more detail).

   More recently, short lengths of sequence from the N- and C-termini of intact proteins, termed "protein terminal sequence tags", have been proposed as attributes for protein identification (Wilkins et al. 1996b; Gooley et al. 1997; Wilkins et al. 1997a). The specificity of protein terminal sequence tags is surprising, especially in organisms with small genomes. For example in *E. coli*, about 60% of proteins have unique N-terminal sequence tags of length 4 amino acids (Fig. 3.3). C-terminal tags of length 4 amino acids are even more specific, being unique to about 90% of all *E. coli* proteins. Where terminal sequence tags are not unique for a protein, relatively few proteins share the same tag. Thus of the 3,008 *E. coli* proteins examined[1] in Fig. 3.3, the most frequent N-terminal tag of length 4 amino acids (MKTL) was found at the start of only 10 proteins, whilst only 4 proteins shared the most frequent C-terminal tag (AKKK). This illustrates that protein terminal sequence tags, when determined analytically for proteins from 2-D gels, should be powerful attributes for identification. However, it must be noted that sequence tags will be less useful in organisms of large proteome size, as there are increasing numbers of proteins that share N- and C-termini. The ideal application for protein sequence

---

[1] All plasmid proteins were excluded here, thus only 3,008 proteins were examined.

```
*************************************************************
*    TagIdent Search on the ExPASy WWW Server              *
*************************************************************
Search performed with following values:

Query name: 2405 sutton
pI =        5.2      Mw =          41578
delta-pI = 0.25      delta-Mw =    415
OS or OC = SAPIENS   KW keyword = ALL


--------------------
7 proteins found
        ACTA_HUMAN   (P03996)
                 pI: 5.24, MW: 41774.62
                 ACTIN, SMOOTH-MUSCLE.

        ACTB_HUMAN   (P02570)
                 pI: 5.29, MW: 41605.54
                 ACTIN, CYTOPLASMIC 1.

        ACTC_HUMAN   (P04270)
                 pI: 5.23, MW: 41784.64
                 ACTIN, ALPHA CARDIAC.

        ACTG_HUMAN   (P02571)
                 pI: 5.31, MW: 41661.65
                 ACTIN, CYTOPLASMIC 2.

        ACTH_HUMAN   (P12718)
                 pI: 5.31, MW: 41642.54
                 ACTIN, GAMMA-ENTERIC SMOOTH-MUSCLE.

        ACTS_HUMAN   (P02568)
                 pI: 5.23, MW: 41816.70
                 ACTIN, ALPHA SKELETAL MUSCLE.

        HIP_HUMAN    (P50502)
                 pI: 5.09, MW: 41324.68
                 HSC70-INTERACTING PROTEIN.
--------------------
```

Fig. 3.2. The identity of a protein can be narrowed down to a few candidates by searching protein databases with protein pI estimated from a gel and protein mass determined by MS techniques. Here data for a human heart protein from Sutton et al. (1997) is matched against human proteins in the SWISS-PROT database using the ExPASy TagIdent tool, with pI ± 0.25 units and mass ± 1%. Only 7 proteins match using these criteria, including the protein identity alpha cardiac actin (ACTC_HUMAN). The use of a further protein attribute in database searching should thus reveal the correct protein identity in this case. It must be noted that this approach may not be applicable where proteins are extensively post-translationally modified as mass values determined by mass spectrometry for a modified peptide may not correspond well with the mass for the unmodified protein as calculated from the database. If a genome is poorly defined, this approach is of limited value as only some of the proteins will be in the database

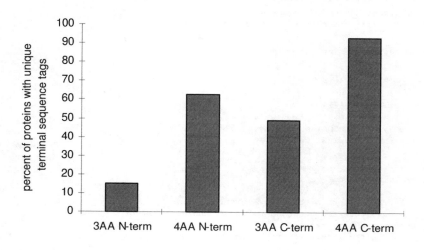

Fig. 3.3. Study of 3,008 proteins from the *E. coli* proteome to determine the number of proteins whose N- and C-termini are unique. Sixty-three percent of proteins have unique N-terminal sequences when their first 4 N-terminal amino acids are considered, and 93% of proteins have unique C-termini when 4 amino acids of C-terminal sequence are considered. Tags of 3 amino acids are less specific, but still unique for many proteins. Data from Wilkins et al. (1997a)

tags is for the identification of proteins from organisms of small proteome size whose genomes are known. In this case, if only one protein within a given pI and mass range is found to carry a certain N- or C-terminal sequence, one can be confident that there is no other, as yet undescribed, protein that could otherwise match the tag.

Most protein terminal sequence tag data to date has been generated by N-terminal Edman degradation of proteins blotted to membranes. In this approach, no treatment of samples is necessary before analysis, sequences are easily assigned, and the estimation of the molar quantity of protein present is possible. Through the combination of rapid cycle times for sequencing, sequenator carousels that accept 16 samples, and parallel processing through analysing high and low abundance spots simultaneously, the maximum throughput is 24 samples per day per machine (Wilkins et al. 1996c; Gooley et al. 1997). If protein samples are blocked or present in subpicomole quantities, Edman degradation cannot generate reliable tag data. However, any membrane-bound sample already subjected to Edman degradation can be subsequently submitted for amino acid analysis (Wilkins et al. 1996c) or C-terminal sequence analysis (Burkhart et al. 1996), thus generating other protein attributes which may be used for identification. In contrast to the ease of assigning N-terminal tags, the generation of C-terminal protein sequence tags remains diffi-

```
***********************************************************
*    TagIdent Search on the ExPASy WWW Server          *
***********************************************************


Search performed with following values:


Query name: LO5


pI =            5.85    Mw =            17625
delta-pI =      0.50    delta-Mw =      3525
OS or OC =      ECOLI   Sequence Tag =  AKEF
Display the N-terminal sequence.


--------------------
108 proteins found
---
Proteins which contain the tag: 2 found
The number before the sequence indicates the position in the
sequence where your tag AKEF has been found (first occurrence).
The sequence tag itself is printed in lowercase.
---
        RBFA_ECOLI        (P09170)
                RIBOSOME-BINDING FACTOR A (P15B PROTEIN).
                pI: 5.96, MW: 15023.29
        1       akefGRPQRVAQEMQKEIALILQREIKDPRLGMMTTVSGV...


        TESA_ECOLI        (P29679)
                ACYL-COA THIOESTERASE I.
                pI: 5.91, MW: 20470.33
        129     ADTLLILGDSLSAGYRMSASAAWPALLNDKWQSKTSVVNA...


---
Proteins which do not contain the tag: 106 found
--------------------
```

Fig. 3.4. Sample TagIdent output for a protein blotted from an *E. coli* 2-D gel (data from Wilkins et al. 1997a). The protein was sequenced for 4 amino acids using Edman degradation to generate an N-terminal protein sequence tag of AKEF. This tag was then matched against the database in conjunction with estimated protein pI and mass. Of the 108 proteins within the specified pI and mass ranges, only two proteins carried the tag anywhere in their sequence. As the sequence tag is found at position 129 in the protein TESA_ECOLI, this protein is unlikely to be the identification of the unknown protein. Only the protein RBFA_ECOLI carries the tag at its N-terminus, thus is highly likely be the identification of the unknown protein

cult. Chemical C-terminal sequencing on membranes generally requires about 100 pmol of protein (Burkhart et al. 1996), and carboxypeptidase digestion, whilst successful for generating sequence in *peptides* via MALDI-TOF MS (Patterson et al. 1995), is yet to be demonstrated for the sequence analysis of significant numbers of whole proteins. A robust and sensitive method for the generation of protein C-terminal sequence data would clearly be useful.

Tools are available on the Internet to match sequence tags against protein databases, in combination with other primary attributes such as protein pI, mass, and species (Table 3.2). A sample output from one of these, the TagIdent program, is shown in Fig. 3.4.

3.3.5 Extensive N-terminal protein sequence

The sequencing of protein N-termini by automated Edman degradation is a technique protein chemists have used for 30 years (Edman and Begg 1967), and whilst there are relatively few proteins whose primary structure has been elucidated by protein sequencing alone, Edman degradation has been the method of choice for protein identification until very recently. As such it should not be ruled out for the identification of proteins from 2-D gels, although there are a few factors to note. Firstly, Edman degradation, by comparison with some other techniques described in Sect. 3.4, is slow. In cases where high repetitive yield, high sensitivity sequencing is desired, sequence is generated at the rate of only 1 amino acid every 40 minutes. Edman degradation also needs relatively large amounts of material, especially when compared to mass spectrometric techniques. But having said that, it should be noted that Edman sequencing of 1 to 5 picomoles of protein is routine, and that in practice most samples prepared for MS analysis that give high quality results are also of this quantity, even though the entire sample may not be consumed during analysis. Finally, Edman degradation requires expensive reagents, with each amino acid of sequence costing US$3–4. The combination of these factors means that extensive Edman degradation of proteins is not suitable where hundreds to thousands of proteins are to be analysed. However, if relatively few proteins from a gel are of interest, or if a protein appears to be unknown by other techniques and cloning its gene is desirable, extensive protein sequencing will be of use. A single amido-black stained spot from a micropreparative 2-D gel will usually provide sufficient material for analysis, and it is possible to generate 40 to 80 amino acids of sequence. This means that some small proteins can be sequenced entirely, from N- to C-terminus, from one 2-D gel sample. As with terminal protein sequence tags, no pretreatment of samples is necessary before Edman analysis, the sequencing itself is automatic and robust, and the sequence results are easily interpreted. If sequence data is to be matched against databases, the program of choice is BLAST (Altschul et al. 1990), which is available at most Internet sites for protein and DNA databases.

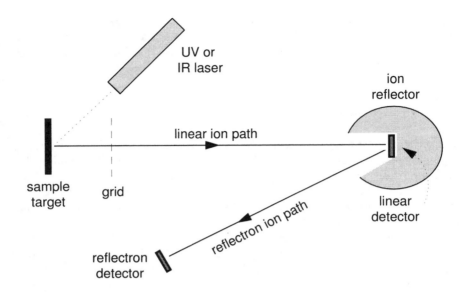

Fig. 3.5. Schematic diagram of a MALDI mass spectrometer. Microlitre quantities of liquid samples are mixed with a matrix molecule, such as α cyano 4 hydroxy cinnamic acid, and dried onto a stainless steel or gold-plated target. A pulsing laser, most commonly a N_2 laser emitting at 337 nm, is then used to irradiate the matrix-embedded sample. This creates molecular ions that are accelerated by an electric field that exists between the sample target and grid. Ions then enter a field-free flight tube with a velocity essentially proportional to their mass, and their time of flight in this tube is measured at the linear detector. To increase mass resolution, some MALDI instruments have an ion reflector. This turns ions around in an electric field, sending them towards a second detector. The reflector makes possible partial peptide sequence analysis by post-source decay (Spengler et al. 1992), as ions whose mass / velocity ratio have changed due to decay during linear flight can be brought into sharp focus by lowering the voltage used in the reflector

3.4 Secondary attributes for protein identification

3.4.1 Attributes of peptides from mass spectrometry

Mass spectrometry is an analytical technique that can sound rather complicated. However, it is simply a technology that weighs any particular molecule, or the component molecules of a mixture. In general, it is not a technique that gives quantitative data but instead, very accurate mass measurements are possible. Mass spectrometers used for the analysis of proteins or peptides can be seen as having two main parts: an ion source which introduces sample into the machine, and an apparatus to measure the mass of the introduced ions. Two types of mass spectrometry are discussed in this chapter. The first is matrix assisted laser desorption / ionisa-

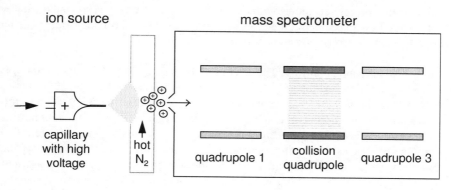

Fig. 3.6. Schematic diagram of ESI-MS, in this case a triple quadrupole mass spectrometer equipped with electrospray ion source. Liquid sample is delivered to the end of the high-potential capillary either under pressure from a pump, or by electrostatic forces alone if via a nanospray capillary (Wilm and Mann 1996). There it forms a mist of small solvent droplets of high electrical charge, containing protein. With the aid of hot nitrogen gas, the solvent evaporates, releasing charged protein ions into the gas phase. These enter a high vacuum environment in the mass spectrometer through a tiny orifice. In the first quadrupole, masses of molecules can be measured. If desired, chosen molecules can then be sent from the first quadrupole to the second, for fragmentation by collision with an inert gas. The mass of fragmentation products can then be measured in the third quadrupole. Note that a further type of mass spectrometer has recently been built that combines an ESI source with a time of flight detector

tion time of flight mass spectrometry (MALDI-TOF MS), which generates ions from solid-phase samples and measures their mass in a flight tube. The second is electrospray ionisation mass spectrometry (ESI-MS), that generates ions from samples in liquid, and measures their mass either in a device called a quadrupole or in a time of flight detector. The principles of each MS technique are illustrated in Fig. 3.5 and 3.6. There is also an ion trap MS available (see Jonscher and Yates 1997) that allows multiple secondary analyses on selected ions, but this will not be discussed here.

Recently, there have been dramatic improvements in mass spectrometry performance due both to improved apparatus design and refinements of sample preparation methods. These are detailed elsewhere in a series of reviews (Wang and Chait 1994; Mann and Talbo 1996; Roepstorff 1997; Jonscher and Yates 1997). In MALDI-TOF MS, the most important advances have been those of the ion reflector and delayed extraction (Brown and Lennon 1995), which together allow isotopic resolution of peptides and very high mass accuracies (Jensen et al. 1996). In ESI-MS, the invention of the capillary nanoelectrospray source (Wilm and Mann 1996) has enabled microlitre amounts of a single sample to be analysed over 30 to 40 minutes, greatly improving the ease of tandem mass spectrometric analysis, especially of peptide mixtures. Both techniques of MALDI-TOF MS and ESI-MS can generate protein attributes with extraordinary sensitivity and precision. Below we outline how mass spectrometry can generate secondary protein attributes using peptide mass fingerprinting and peptide fragmentation.

Peptide mass fingerprinting. The technique of peptide mass fingerprinting was independently described in 1993 by a number of groups (Henzel et al. 1993; James et al. 1993; Mann et al. 1993; Pappin et al. 1993; Yates et al. 1993). It can be summarised as follows. As a first step, proteins separated by 2-D gel electrophoresis are digested either in-gel or *in situ* on membranes with an enzyme or reagent that specifically cleaves at certain amino acids. A commonly used enzyme is trypsin, that cleaves only at the C-terminal side of arginine or lysine. The exact masses of peptides that result from this cleavage are then measured in a mass spectrometer, either by MALDI-TOF MS or ESI-MS techniques. Either technique can achieve peptide mass accuracies better than 0.1 mass units. All peptide masses are finally matched against theoretical peptide masses for proteins in databases, that have been generated by "cleaving" proteins in the manner of the experimentally used enzyme. The matching output is a list of proteins ranked by the number of peptides shared with the unknown protein, where the correct identification for an unknown protein is likely to be that with the largest number of peptide "hits". Confidence in identification is achieved by looking for a large gap in the number of matching peptides between the top and second ranked protein, and a good coverage of the top-ranked protein with the experimentally determined peptides.

The spectrometric technique of choice for measuring peptide masses is a subject of debate, however MALDI-TOF MS is emerging as the best alternative for rapid screening. There are many reasons for this. Firstly, the mass accuracy and resolution of reflectron delayed extraction MALDI-TOF MS for peptides now rivals that of ESI-MS, allowing very stringent mass tolerance to be used in database searching (Jensen et al. 1996). Secondly, MALDI-TOF MS is relatively insensitive to contamination from buffer components such as salts and detergents, which allows peptides taken from in-gel or on-membrane digests to be analysed directly. If necessary, samples can easily be washed "on target" (Vorm et al. 1994). Thirdly, most MALDI machines can now accommodate targets with many samples and automatic sample analysis is being developed, allowing high throughput. For example, Shevchenko et al. (1996) automatically generated high-quality spectra for 65 different proteins in 6 hours. Fourthly, MALDI-TOF MS can routinely analyse peptides at the 50 to 100 fmol level, such that a fraction of a coomassie-blue stained spot from a gel is sufficient material for analysis. Finally, MALDI-TOF MS data is relatively easy to interpret, as most peptides carry only one charge and have only one peak in a spectrum. This contrasts to ESI-MS where many tryptic peptides are present as doubly charged ions which must be deconvoluted to find their true mass. However, the modern quadrupole mass spectrometer has sufficient mass resolution that multiply charged ions are not difficult to deconvolute.

There are a number of programs available on the Internet for protein identification by peptide mass fingerprinting (Table 3.2). These generally accept primary protein attributes such as species and intact protein mass, as well as the secondary attribute of peptide masses. In addition, users must specify enzyme used, peptide mass tolerance, if missed enzymatic cleavages are allowed, and if modifications have been made to cysteine residues (e.g. alkylation by DTT or acrylamide). An output from one identification program is shown in Fig. 3.7. It is interesting to note that identification programs have also been installed locally on some computers operating MALDI-TOF MS machines, such that peptide mass data can be automatically matched against databases (Shevchenko et al. 1996). This type of approach greatly assists high throughput protein identification.

Protein identification by peptide mass fingerprinting is becoming a widely used technique for proteins from 2-D gels, and can yield spectacular results. For example, of 150 proteins analysed from 2-D separations of *S. cerevisiae*, 80% were identified without ambiguity by MALDI-TOF MS mass fingerprinting alone (Shevchenko et al. 1996). However, many users have faced a lower rate of identification, and have described methods that combine peptide mass data with other protein attributes in order to achieve high-confidence identification. These include the combination of peptide mass data with partial Edman sequence (Rasmussen et al. 1994; Patterson et al. 1996), peptide masses with protein amino acid composition (Cordwell et al. 1995; Wheeler et al. 1996), peptide mass data from one digestion with mass data from digestion of a duplicate sample with a second enzyme (Pappin et al. 1993), and peptide mass data with that from peptide fragmentation (see below). Numerous programs, including MS-Edman, Mowse, MassSearch and MultiIdent can achieve protein identification by matching different combinations of protein attributes against databases, and are available on the Internet (Table 3.2).

Peptide mass data, as well as being an attribute for protein identification, can represent a starting point for the examination of protein modifications or processing. Peptides from a protein that do not match those from the database may carry post-translational or artifactual modifications, or may result from protein N- or C-terminal truncation (for detail see Chap. 4; Burlingame 1996; Roepstorff 1997). However, it must be noted that peptide mass fingerprinting will rarely yield all peptides from a particular protein. This is because some peptides, especially those that are large and/or very hydrophobic, are either not extracted or not quantitatively extracted from a gel or blot, and others are not ionised efficiently during mass spectrometry. Thus peptide mass fingerprinting is not commonly used for the definition of protein N- or C-termini in proteome projects, and is a technique unsuitable for the quantitation of protein present in any spot on a gel.

Peptide fragmentation to generate partial sequence. In peptide mass fingerprinting, peptides of proteins generated by enzymatic or chemical digestion are characterised by a single attribute, this being their mass. This attribute is valuable for protein identification in the context of other peptide mass values from the same protein. However, by itself it reveals little about the peptide or the protein from which it was derived. In order to generate further attributes for a peptide that can be applied to protein identification, a series of MS methods have been described to

Search result

One match was found.

Peptides matched	Mass [kDa]	Database accession	Protein Name
9	14.284	Swissprot P33633	YFID_ECOLI HYPOTHETICAL 14.3 KD

Match results

Cysteine is Cys carbamidomethyl. Methionine is unmodified/native.

Measured [Da]	Calculated [Da]	Mass is	Diff. [Da]	Start residue	End residue	Sequence
432.291	432.221	mono	0.07	27	30	(K)GEAR(C)
510.632	510.268	mono	0.364	89	92	(K)HPEK(Y)
590.421	590.334	mono	0.087	31	35	(R)CIVAK(A)
865.631	865.442	mono	0.189	49	55	(K)LGDIEYR(E)
890.701	890.51	mono	0.191	93	99	(K)YPQLTIR(V)
1281.011	1280.721	mono	0.289	56	66	(R)EVPVEVKPEVR(V)
1382.031	1381.759	mono	0.271	89	99	(K)HPEKYPQLTIR(V)
1435.101	1434.782	mono	0.319	67	79	(R)VEGGQHLNVNVLR(R)

Cysteine is Cys carbamidomethyl. Methionine is oxidized.

Measured [Da]	Calculated [Da]	Mass is	Diff. [Da]	Start residue	End residue	Sequence
1020.771	1020.576	mono	0.194	1	9	( )MITGIQITK(A)

These 9 peptides cover 60 out of 127 amino acids (47.2 %).
5 masses (protonated) were not identified : 731.381 782.621 1061.761 2365.541 2397.531

Fig. 3.7. Output from the PeptideSearch program of Mann and colleagues for peptide mass fingerprinting protein identification. A protein from an amido black stained PVDF blot of an *E. coli* 2-D gel was digested with trypsin, and peptide masses measured by MALDI-TOF MS in reflectron delayed extraction mode with external calibration. Peptides were matched against the SWISS-PROT database with mass tolerance of ± 0.5 Da, for proteins between 10 and 20 kDa from all species. One protein was found with more than 5 peptide matches. Note the good agreement of measured and calculated peptide masses and that one peptide was found to carry an oxidized methionine. Bracketed letters in the peptide "Sequence" column are the amino acids that flank a peptide in the protein; these amino acids are for reference only and are not part of the peptide mass itself

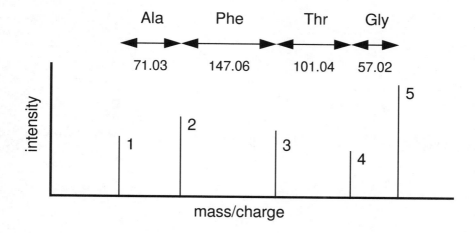

Fig. 3.8. Peptide fragmentation is most useful when it produces a ladder of peptides, where the mass differences between each peptide (arrows and values) correspond exactly to that of a certain amino acid. In this manner, partial sequence of a peptide can be "read". In this schematic diagram, a sequence of AFTG is present. Peptide 5 would have been the peptide XXAFTG, peptide 4 XXAFT, peptide 3 XXAF, peptide 2 XXA and peptide 1 XX

fragment peptides in different ways. Sequential fragmentation of peptides can be done enzymatically or chemically, removing amino acids one by one from the N- or C-terminus of a peptide to generate a ladder of peptides. Alternatively, peptides can be fragmented within mass spectrometers using post-source decay (PSD) or collision-induced dissociation (CID) techniques in MALDI and ESI machines respectively. In all methods, the objective is to produce spectra containing a series of peptide peaks differing by the mass of amino acid residues, thus allowing stretches of peptide sequence to be deduced (Fig. 3.8). These fragmentation approaches are discussed briefly below.

Peptide ladder sequencing can be undertaken in two ways. Chait et al. (1993), who first defined the concept, subjected peptides to N-terminal degradation in a controlled chemical fashion. For each cycle of the chemistry, the N-terminal amino acid was removed from most copies of a peptide, but a small percentage of copies were chemically "blocked" to make it impossible to remove amino acids in subsequent cycles. The chemistry was then repeated to generate a series of peptides of different size, determined by the number of amino acids that had been removed from their N-terminus. The masses of peptides were finally measured by MALDI-TOF MS, to yield a peptide ladder (see Fig. 3.8). A volatile reagent was subsequently described elsewhere to facilitate the process of N-terminal degradation in this approach (Bartlet-Jones et al. 1994). A second approach for generating peptide ladders involves the use of carboxypeptidase enzymes. This generates a series of peptides which differ in size according to the number of amino acids that have been

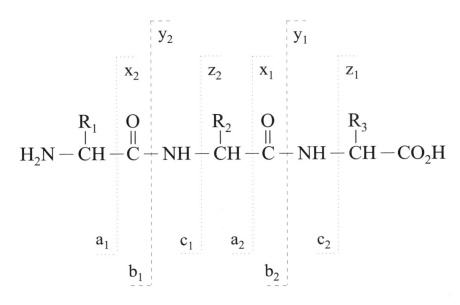

Fig. 3.9. Common fragment ions from peptides and their notation (after Biemann 1990). Here, it is shown how a peptide of 3 amino acids can fragment into a series of daughter ions. Fragmentation at the peptide bonds will produce b series ions if the peptide positive charge remains at the peptide N-terminus, or y series ions if the charge remains at the peptide C-terminus. Only the charged portion of the peptide will be detected after fragmentation. The b and y ion fragmentation can produce a ladder of peptides whose mass differences give sequence information according to which amino acid has been added or lost (see Fig. 3.8). Other fragmentation can also occur along the peptide backbone, to generate a and c ions, or x and z ions, again depending if the charge is localised at the peptide N- or C-terminus after the peptide is broken

removed from their C-terminus. In this case, MALDI-TOF MS presents what is probably the simplest means of measuring the peptide masses (Thiede et al. 1995; Patterson et al. 1995), even though other forms of mass spectrometry have also been used. Both chemical and enzymatic means of generating peptide ladders can potentially generate relatively long stretches of sequence (about 30 residues). However very high mass accuracy is required to obtain unambiguous sequence, especially to distinguish between the residues of lysine (mass 128.09) and glutamine (mass 128.06). Isoleucine and leucine will not be distinguishable as they are of identical mass, but they can be distinguished by different mass profiles during their degradation.

The analysis of peptide post-source decay (PSD) fragments in reflectron MALDI mass spectrometers can yield partial sequence information (Spengler et al. 1992). To undertake this analysis, peptide mass fingerprinting is carried out as per normal, and after masses of peptides are established, a single peptide of interest is electrically selected on the mass spectrometer as a "parent ion" for further analysis. Conveniently, a single sample in the MALDI-TOF MS can be used sequentially for this. The parent ion is then desorbed and accelerated in the MALDI apparatus as per usual (see Fig. 3.5), and any parent ion fragments, which have decomposed during the flight to the ion reflector, are measured by lowering the voltage in the reflector from its normal value, thus sending fragments of various sizes to the reflectron detector. The manner in which peptides fragment is now well understood (Fig. 3.9), however MALDI-PSD often produces incomplete fragmentation, making the *de novo* interpretation of sequence difficult or impossible. Nevertheless, protein identification programs now exist that can match mass values from uninterpreted PSD spectra, in conjunction with other attributes such as species, protein mass, and parent ion mass, against similar values in databases (Table 3.2) (Griffin et al. 1995). An output from one such program is shown in Fig. 3.10, illustrating that even incomplete fragmentation data of one peptide from a protein can generate attributes of sufficient specificity to identify a protein unambiguously. Note that MALDI-PSD is not used extensively for peptide sequencing as the fragmentation of peptides is inconsistent, with many peptides showing little or no decomposition.

The method of choice for peptide fragmentation and sequencing is currently collision-induced dissociation (CID), which is done by triple quadrupole ESI-MS (Fig. 3.6) or recently available MALDI-TOF instruments equipped with a collision cell. In ESI-MS, masses of peptide ions introduced via the electrospray source are measured in the first quadrupole of the mass spectrometer. One peptide of interest can then be sent to the second quadrupole, where it is fragmented by collision with an inert gas. The masses of fragmentation products can then be measured in the third quadrupole. By contrast with MALDI-PSD fragmentation, CID is robust and general, frequently producing a series of y ions from which sequence can be deduced (Fig. 3.9). One spectacular application of this to date is the sequencing of 73 residues from a 45 kDa protein, from a series of tryptic peptides of length up to 16 amino acids (Wilm et al. 1996). The starting material for analysis was about 200 ng. The manual interpretation of spectra from CID remains a challenge, however there are computer programs that can automatically interpret these (e.g. Yates et al. 1995). Yet, for the purpose of protein identification, complete interpretation of the spectra is in most cases unnecessary. A few residues of sequence obtained from a CID spectra, called a "peptide sequence tag" (Mann and Wilm 1994), combined with the peptide parent ion mass and the distance in mass units to the N- and C-termini of the peptide, will in most cases be sufficiently specific to identify a protein. This is convincingly shown in Shevchenko et al. (1996) where all of 49 proteins analysed yielded an identification. The reason why peptide sequence tags are not always used as a first step in identification is that compared to peptide mass fingerprinting on MALDI-TOF MS the approach is relatively slow, needs expert operators, and requires manual manipulations and data analysis. However, the future possibilities for protein identification and even sequencing of proteins by CID via ESI-MS are tremendous. The recent introduction of the electrospray quadrupole time of flight (QTOF) instrument also seems likely to simplify the *de novo* sequencing of peptides.

Sample ID (comment): coli 46
Database searched: SwissProt.r34
Digest used: Trypsin
Ion Types Considered: a b B y n h I

Search mode:	Max. # of unmatched ions:	Max. # of missed cleavages:	Cysteine Modification:
no errors	1	1	unmodified

Parent mass: 890.7500 (+/- 0.50 Da)
Fragment Ions used in search:
198.2, 225.7, 232.8, 260.9, 271.1, 288.2, 371.8, 388.9, 422.1, 440.3, 585.2, 603.5 (+/-0.50 Da)
Composition Ions present: [RP] [LI] [KQ]

Rank	MS-Digest Index #	SwissProt Accession	Protein MW (Da)	Species	Calculated MH+ (Da)	MH+ Error (Da)	Sequence
1	53701	P33633	14284.3	ECOLI	890.5100	0.240	(K)YPQLTIR(V)

198.2	225.7	232.8	260.9	271.1	288.2	371.8	388.9	422.1	440.3	585.2	603.5
PQ-28	PQ	a2	b2	y2-NH3	y2	b3-NH3 y3-NH3	b3	y3	PQLT-H2O	PQLT	b5-H2O	b5

Fig. 3.10. Output from the MS-TAG program by Karl Clauser and Peter Baker (Table 3.2) for protein identification with peptide ion fragment data. A protein from an amido black stained PVDF blot of an *E. coli* 2-D gel was digested with trypsin, and peptide masses measured on a MALDI-TOF MS in reflectron delayed extraction mode with external calibration (see Fig. 3.5). The peptide of mass 890.7 was then subjected to post-source decay fragmentation. Fragment masses less than 150 Da were assigned as immonium ions, and all other fragments were entered into the program uninterpreted. Matching was done with all *E. coli* proteins in the SWISS-PROT database between 10 kDa to 20 kDa, with a mass tolerance of ± 0.5 Da for parent and fragment ions. One protein was found with a peptide matching the entered fragmentation data. This was protein P33633 (YFID_ECOLI), which confirmed the identification from peptide mass fingerprinting alone (Fig. 3.7). The final table in this program output gives details of how the PSD ions correspond to the theoretical fragments of the peptide. A complete b or y ion series was not found here. The novice user may find this figure daunting!

3.4.2 Protein amino acid composition

Protein amino acid composition is determined by the numbers of each amino acid present in a molecule, and is usually described in percentages. It is an attribute that is sequence independent, and thus different to protein attributes such as peptide masses or sequence tag (Hobohm and Sander 1995). Identification of proteins by amino acid composition is, in principle, straightforward. A protein from a 2-D gel has its amino acid composition determined, and this composition, expressed as ratios or as a percent, is then compared to theoretical compositions of proteins in databases. A score, representing the numerical difference between two compositions, is used to rank proteins in the database from best-matching to worst-matching. Proteins ranked at the top of the list, having compositions closest to the unknown protein, are then examined to determine if they represent a correct identification.

Technically, the amino acid composition of proteins from 2-D gels can be determined in two ways. Latter et al. (1984) who first described the identification of 2-D separated proteins by composition alone, determined composition by radiolabelling proteins with one radioactive amino acid at a time, and determining the ratios between amino acids by comparing quantitative densitometric measurements from different gels. More recently this approach has been extended such that proteins are radiolabelled with two amino acids at a time, where one amino acid carries ^{3}H, ^{14}C or ^{35}S, and the second amino acid carries a different radiolabel to the first. After 2-D gel electrophoresis, the ratios between amino acids in a protein are then established via their different radioactivities, to provide compositional information (Garrels et al. 1994; Maillet et al. 1996). These approaches have the advantage of producing data in parallel for up to hundreds of spots for any sample. However, for reasons most likely due to difficulties of sample preparation and data interpretation, they have not emerged as general, high throughput methods.

Alternatively, numerous groups have generated amino acid composition data by hydrolysing proteins with strong acid, and then analysing the resulting free amino acids with chromatographic techniques (Eckerskorn et al. 1988; Jungblut et al. 1992; Shaw 1993; Hobohm et al. 1994; Galat et al. 1995; Wilkins et al. 1996a). Generally protein samples are from micropreparative 2-D gels blotted to PVDF membranes, allowing extensive washing of samples prior to analysis. At its most streamlined, PVDF-bound proteins can be hydrolysed in 1 hour, amino acids extracted in one step, and each sample then automatically derivatised and chromatographically separated in about 40 minutes (Ou et al. 1996; Yan et al. 1996). This gives a potential throughput of 20 samples per day per machine. A lower limit of 250 ng of protein, representing a medium-sized coomassie stained spot on a gel, is required for the method. As there is no requirement for radiolabelling this approach is generally applicable. It also quantitates the amount of protein present in a protein spot in molar or gram terms. Another feature of chromatographic amino acid analysis is that it can be undertaken on PVDF-bound protein samples after they have been subjected to Edman degradation sequence tagging or monosaccharide analysis (see Chap. 4; Packer et al. 1996; Wilkins et al. 1996c). Such multiple analyses can maximise the amount of information that can be obtained from a single sample of protein.

Two programs exist on the Internet that can use protein amino acid composition for identification purposes (Table 3.2). The output from one of these, AAComp-

Ident, is shown in Fig. 3.11. The confidence that one has in a top-ranked protein representing a correct identity can be addressed in two ways. Firstly, by examining the numerical score, which represents the goodness-of-fit of the top ranked protein with a protein in the database (Hobohm et al. 1994; Shaw 1993). In this manner, a protein with a score passing a certain threshold is considered to be identified with confidence. Secondly, the numerical score for the top-ranked protein as well as the difference in score between the top-ranked protein and the second-ranked protein can be considered (Wilkins et al. 1996a). In this case, the goodness-of-fit and the uniqueness of fit are assessed, and only top ranking proteins that show a suitable score pattern are considered to be identified with confidence. As may be expected, the incorporation of pI and mass attributes into database searching greatly increases the uniqueness of fit and thus confidence of identification that can be obtained (Fig. 3.11). In ideal conditions, about 45% of proteins from species with sequenced genomes can be identified with high confidence by composition, when used with attributes of protein species, pI and mass (Wilkins et al. 1996a). However, this requires that samples are separated to purity by electrophoretic techniques, and that there is minimal contamination with amino acids such as glycine from the gel itself. When identification is not achieved by composition data alone, this data can be combined with other attributes such as protein N-terminal sequence tag, or peptide mass data to achieve high-confidence identification (Cordwell et al. 1995; Wilkins et al. 1996c). These approaches exploit the fact that if proteins are not confidently identified as judged by score patterns, the correct protein identity is nevertheless frequently at the top, or close to the top of the list of best-matching proteins. The programs AAComplIdent and MultiIdent (Table 3.2) can accept combinations of protein amino acid composition data with protein sequence tags and peptide mass data respectively.

3.5 Cross-species protein identification

Protein sequence databases are growing at a near-exponential rate, and whilst this is generating massive amounts of information for some model organisms, the majority of species remain more or less molecularly undefined. The magnitude of this is illustrated in the SWISS-PROT database (release 34), where 38% of all sequence comes from just 10 organisms (see Sect. 5.2.1). The question that then arises is: can the genome / proteome data from model organisms be used for the identification and understanding of proteins from other species? In other words, can an unknown protein from a poorly defined species, for example *Serratia marcescens*, be confidently identified by comparing its attributes with proteins from a well-defined species, such as *E. coli*? Given the evolutionary relatedness of protein sequences between species, the answer is, of course, yes. However, as discussed in this chapter, protein identification in proteome projects rarely uses extensive pro-

```
***********************************************************
*    Amino Acid Composition Search on the ExPASy WWW Server    *
***********************************************************
Species searched: ESCHERICHIA
Keyword searched: ALL
Tagging: MQVI


SpotNb LQE
==========

pI:   6.38        Range: (  5.88,   6.88)
Mw:  14132        Range: (  9892,  18372)


The closest SWISS-PROT entries (in terms of AA composition)
for the species ESCHERICHIA and the specified keyword:
Rank Score    Protein     (pI       Mw)  N-terminal Sequence
=========================================================

*  1    21   RL9_ECOLI     6.17     15769  mqviLLDKVANLGSLGDQVNVKAGYARNFLVPQGKAVPAT
   2    28   MTLD_ECOLI    5.36     41139  MKALHFGAGNIGRGFIGKLLADAGIQLTFADVNQVVLDAL
   3    31   TPX_ECOLI     4.75     17764  SQTVHFQGNPVTVANSIPQAGSKAQTFTLVAKDLSDVTLG
   4    33   RFFD_ECOLI    5.22     45839  MSFATISVIGLGYIGLPTAAAFASRQKQVIGVDINQHAVD
   5    36   GAL1_ECOLI    5.28     41311  SLKEKTQSLFANAFGYPATHTIQAPGRVNLIGEHTDYNDG


The closest SWISS-PROT entries (in terms of AA composition)
for the specified keyword and any species:
Rank Score    Protein     (pI       Mw)  N-terminal Sequence
=========================================================

*  1    21   RL9_ECOLI     6.17     15769  mqviLLDKVANLGSLGDQVNVKAGYARNFLVPQGKAVPAT
   2    22   TPIS_ARATH    5.24     27155  MARKFFVGGNWKCNGTAEEVKKIVNTLNEAQVPSQDVVEV
   3    22   DHAS_CORFL    4.75     36169  MTTIAVVGATGQVGQVMRTLLEERNFPADTVRFFASPRSA
   4    23   DHAS_CORGL    4.79     36212  MTTIAVVGATGQVGQVMRTLLEERNFPADTVRFFASPRSA
   5    23   AKAB_CORFL    4.63     44797  MALVVQKYGGSSLESAERIRNVAERIVATKKAGNDVVVVC


The SWISS-PROT entries having pI and Mw values in the specified range
for the specified species and keyword:
Rank Score    Protein     (pI       Mw)  N-terminal Sequence
=========================================================

*  1    21   RL9_ECOLI     6.17     15769  mqviLLDKVANLGSLGDQVNVKAGYARNFLVPQGKAVPAT
   2    59   YHAR_ECOLI    6.09     16348  MNSNRMLFHVPLPRRSDAGGCMKKIIETQRAPGAIGPYVQ
   3    63   PUR6_ECOLI    6.03     17649  SSRNNPARVAIVMGSKSDWATMQFAAEIFEILNVPHHVEV
   4    71   HOLC_ECOLI    5.91     16633  MKNATFYLLDNDTTVDGLSAVEQLVCEIAAERWRSGKRVL
   5    80   RPIB_ECOLI    6.58     16073  MKKIAFGCDHVGFILKHEIVAHLVERGVEVIDKGTWSSER
```

Fig. 3.11. Output from the ExPASy AACompIdent program, for protein identification with protein amino acid composition, and in this case, protein sequence tag. A single PVDF-bound protein spot from an *E. coli* 2-D gel was submitted for Edman degradation, which generated a 4 amino acid sequence tag of MQVI. The same protein sample was then used for chromatographic amino acid analysis (Wilkins et al. 1996c). Protein attributes were matched against the SWISS-PROT database to generate these 3 lists of best-matching proteins. The first list matched against all 3606 *E. coli* proteins in the SWISS-PROT database (rel. 34). The top ranked protein by amino acid composition carried the sequence tag at its N-terminus, thus indicating RL9_ECOLI as the protein identity (bold). The second list matched amino acid composition against all 59021 proteins from all species in the database, and showed that the empirical composition data for this protein was sufficiently unique to rank the correct identity first. The third list matched compositional data only against *E. coli* proteins in the specified pI and mass range. Notable is the large score difference between the top and second-ranked protein here, which provides extra confidence that the top-ranked protein is the correct identification

tein sequence as it is expensive and/or slow to generate. So, we must instead ask: how are the more commonly used protein attributes of pI, intact mass, peptide masses, amino acid composition, and protein sequence tags conserved across species boundaries? Or more precisely, what is the effect of cross-species sequence divergence on these attributes?

A theoretical study of 65 cross-species comparisons involving 21 different types of protein made some clear, but perhaps surprising findings on the cross-species conservation of protein attributes (Wilkins and Williams 1997). Protein pI was found to be poorly conserved, with some proteins showing as much as 2 pI units difference across species boundaries. By comparison, the intact mass of proteins was well conserved, with a mean absolute difference of 1.9%. Peptide masses were not well conserved across species boundaries, with few or no peptides being conserved when sequence identity between two proteins was below 75%. However, protein amino acid composition was well conserved across species boundaries, with many proteins not showing large compositional differences between species until sequence identity was below 60%. The poor conservation of peptide mass data is expected, as a single amino acid substitution in any peptide can drastically change its mass. However, the reasons why protein mass and composition are well conserved yet protein pI is not remain undefined, although we speculate that this concerns 3-D structural constraints. Nevertheless, the theoretical findings concerning conservation of composition and peptide masses were reflected experimentally elsewhere (Cordwell et al. 1995; Wasinger et al. 1995) where a total of 13 of 17 proteins were correctly identified cross-species with compositional data, but only 2 of the same proteins identified correctly with peptide mass fingerprinting.

In practice, cross-species protein identification should be applicable to the study of proteins from many organisms, and is likely to be most useful for so-called housekeeping proteins, which are generally well conserved. If high throughput analytical techniques are to be used, it may be necessary to consider many protein attributes to ensure confident protein identification. This will particularly be the case if matching is to be done across large phylogenetic distances. Perhaps the most comprehensive program currently available for cross-species identification is the ExPASy MultiIdent tool (Table 3.2), which can accept attributes of protein pI, mass, composition, peptide masses, and sequence tag, and can match against any proteins from a desired species, genus, family or kingdom. An output from this program is shown in Fig. 3.12. Alternatively, the MS-Tag program (Table 3.2) has an "allow errors mode" which can simulate the effects of single amino acid substitutions in peptides from databases to achieve a better fit with experimental mass spectrometric peptide fragmentation data. It is anticipated that with increasing availability of genome / proteome data from many small and large organisms, cross-species protein identification will become of increasing interest in the near future.

```
****************************************************************************
*    MultiIdent Protein Identification on the ExPASy WWW Server            *
****************************************************************************
Species searched: ALL
Keyword: ALL
Tag: No_Tagging
The enzyme is Trypsin.
Cysteine in reduced form.
Peptide Mass Tolerance: 2 Dalton
Using average masses of the occurring amino acid residues
and interpreting your peptide masses as [M+H]+.


SpotNb DOG3
===========

pI:    4.60        Range: (  3.60,   5.60)
Mw:   33500        Range: ( 26800,  40200)


AA COMPOSITION
The closest SWISS-PROT entries having pI and Mw values
in the specified range for the specified species and keyword:
Rank Score    Protein    (pI      Mw)  Description
================================================================
   1     7  TPMA_RAT     4.71   32695 TROPOMYOSIN ALPHA CHAIN, SKELETAL MUSCLE
   2     7  TPMA_COTJA   4.70   32766 TROPOMYOSIN ALPHA CHAIN, SKELETAL MUSCLE
   3     7  TPMA_RABIT   4.69   32681 TROPOMYOSIN ALPHA CHAIN, SKELETAL AND
   4     9  TPMA_XENLA   4.69   32651 TROPOMYOSIN ALPHA CHAIN, SKELETAL MUSCLE
   5     9  TPM1_HUMAN   4.69   32709 TROPOMYOSIN ALPHA CHAIN, SKELETAL MUSCLE


PEPTIDE-MASS FINGERPRINTING
The best-matching set of peptides for SWISS-PROT entries for the species ALL and
keyword ALL, having pI and Mw values in the specified range:
Rank Hitrate  Protein      Modification      Matched Peptide-masses
================================================================================
   1  0.235    TPMA_RAT                     1539.64 /  926.96 /  861.88 /  722.78
   2  0.235    TPMA_COTJA                   1539.64 / 1286.47 /  861.88 /  722.78
   3  0.235    TPMA_RABIT                   1539.64 /  926.96 /  861.88 /  722.78
   4  0.235    TPMA_XENLA                   1539.64 / 1286.47 /  861.88 /  722.78
   5  0.235    TPM1_HUMAN                   1539.64 /  926.96 /  861.88 /  722.78


INTEGRATED SCORES
Rank  Int_score Protein       Modification
==========================================
   1      5.4     TPMA_RAT
   2      5.4     TPMA_COTJA
   3      5.4     TPMA_RABIT
   4      6.9     TPMA_XENLA
   5      6.9     TPM1_HUMAN
```

Fig. 3.12. Output from the ExPASy MultiIdent program (Wilkins et al., unpublished), illustrating cross-species protein identification. A protein from a 2-D separation of dog heart was subjected to amino acid analysis and a duplicate spot used for peptide mass fingerprinting. Data was matched against all proteins in the SWISS-PROT database with pI ± 1 units, and apparent mass ± 20%. The first list of matches shows the 5 top ranked proteins by composition within the pI and mass ranges. Tropomyosins from rat, quail, rabbit, frog and human were found to be closest to the unknown dog protein. The second list shows the result of matching experimental peptide mass data against the 500 best-matched proteins by composition. The top-ranked proteins were all tropomyosins, sharing 4 peptides with the dog protein. The third list shows an integrated matching where protein pI, mass, composition, and peptide masses are considered simultaneously. The protein identification is clearly tropomyosin

3.6 Future developments and conclusions

Protein identification will continue to play an important role in the understanding of biological systems for many years, especially with continuing advances in genome projects. Because of this, it is anticipated that further effort will be directed towards the development of increasingly affordable, rapid, and automated identification technologies. The recently described ion trap and QTOF (quadrupole time of flight) mass spectrometers will doubtless play a role for the routine and rapid *de novo* sequencing of peptides. The development of novel parallel identification strategies does not currently seem to be a major objective, even though they should offer considerable increases on current throughput. However, one can imagine that the excellent tolerance to complexity of ESI-MS could be further exploited to routinely allow the identification of more than one protein at a time. One can equally imagine that the blotting of a 2-D gel to a membrane, followed by a global digestion to peptides, and finally the automatic scanning of this membrane in a MALDI-TOF MS could certainly save much time in sample preparation. This technology may not be far away.

As a final word, it should be noted that protein identification, as described in this chapter, has in the most part involved the matching of analytically-gained protein attributes against sequences in databases predicted from the translation of genomic or complementary DNA. Thus whilst "confident" protein identification can be achieved in many ways, this does not necessarily mean that the best-matching amino acid sequence in the database exactly represents the protein under study. Isoforms of proteins, which can be widespread in some protein types, may go undetected. Protein N- and C-termini, which define if or how a protein was processed, are frequently not defined. Variations in mRNA splicing, which may change a few or many amino acids in a protein, are also likely to be overlooked. And last but by no means least, errors in databases may go unchecked. Some software is beginning to consider these possibilities in order to better understand analytical results (Wilkins et al. 1997b), however comprehensive means of assessing all the above issues are yet to be established. These issues will no doubt be addressed in the near future.

Acknowledgments

We are grateful to Vassiliki Kordorouba, Dr Colin Wheeler and Nadia Wermeille for providing the protein data used in some figures. MRW acknowledges support from the Helmut Horten Foundation. AAG acknowledges the support of the Australian Proteome Analysis Facility, Macquarie University Centre for Analytical Biotechnology, the Australian Medical Research Council, and the Australia Research Council.

References

Aebersold R, Patterson SD (1997) Current problems and technical solutions in protein biochemistry. In: Angeletti RH (ed) Protein structure. Academic Press, in press

Altschul SF, Gish W, Miller W, Myers EW, Lipman DJ (1990) Basic local alignment search tool. J Mol Biol 215:403–410

Bartlet-Jones M, Jeffery WA, Hansen HF, Pappin DJ (1994) Peptide ladder sequencing by mass spectrometry using a novel, volatile degradation reagent. Rapid Commun Mass Spectrom 8:737–742

Biemann K (1990) Applications of tandem mass spectrometry to peptide and protein structure. In: Burlingame AL, McCloskey JA (eds) Biological Mass Spectrometry. Elsevier, Amsterdam, p 176

Bjellqvist B, Hughes G, Pasquali C, Paquet N, Ravier F, Sanchez JC, Frutiger S, Hochstrasser D (1993) The focusing positions of polypeptides in immobilized pH gradients can be predicted from their amino acid sequences. Electrophoresis 14:1023–1031

Bjellqvist B, Basse B, Olsen E, Celis JE (1994) Reference points for comparisons of two-dimensional maps of proteins from different human cell types defined in a pH scale where isoelectric points correlate with polypeptide compositions. Electrophoresis 15:529–539

Brown RS, Lennon JJ (1995) Mass resolution improvement by incorporation of pulsed ion extraction / ionisation linear time-of-flight mass spectrometry. Anal Chem 67:1998–2003

Burkhart WA, Moyer MB, Bailey JM, Miller CG (1996) Electroblotting proteins to teflon tape and membranes for N- and C-terminal sequence analysis. Anal Biochem 236:364–367

Burlingame AL (1996) Characterisation of protein glycosylation by mass spectrometry. Curr Opin Biotechnol 7:4–10

Chait BT, Wang R, Beavis RC, Kent SB (1993) Protein ladder sequencing. Science 262:89–92

Cordwell SJ, Wilkins MR, Cerpa-Poljak A, Gooley AA, Duncan M, Williams KL, Humphery-Smith I (1995) Cross-species identification of proteins separated by two-dimensional gel electrophoresis using matrix-assisted laser desorption time of flight mass spectrometry and amino acid composition. Electrophoresis 16:438–443

Eckerskorn C, Jungblut P, Mewes W, Klose J, Lottspeich F (1988) Identification of mouse brain proteins after two-dimensional electrophoresis and electroblotting by microsequence analysis and amino acid composition. Electrophoresis 9:830–838

Eckerskorn C, Strupat K, Schleuder D, Hochstrasser DF, Sanchez JC, Lottspeich F, Hillenkamp F (1997) Analysis of proteins by direct scanning infrared MALDI mass spectrometry after 2D-PAGE separation and electroblotting. Anal Chem, in press

Edman P, Begg G (1967) A protein sequenator. Eur J Biochem 1:80–91

Galat A, Bouet F, Rivière S (1995) Amino acid compositions of proteins and their identities. Electrophoresis 16:1095–1103

Garrels JI, Futcher B, Kobayashi R, Latter GI, Schwender B, Volpe T, Warner JR, McLaughlin CS (1994) Protein identification for a *Saccharomyces cerevisiae* protein database. Electrophoresis 15:1466–1486

Gooley AA, Ou K, Russell J, Wilkins MR, Sanchez JC, Hochstrasser DF, Williams KL (1997) A role for Edman degradation in proteome studies. Electrophoresis 18, in press

Griffin PR, MacCoss MJ, Eng JK, Blevins RA, Aaronson JS, Yates JR 3rd (1995) Direct database searching with MALDI-PSD spectra of peptides. Rapid Commun Mass Spectrom 9:1546–1551

Henzel WJ, Billeci TM, Stults JT, Wong SC, Grimley C, Watanabe C (1993) Identifying proteins from two-dimensional gels by molecular mass searching of peptide fragments in protein sequence databases. Proc Natl Acad Sci USA 90:5011–5015

Hobohm U, Sander C (1995) A sequence property approach to searching protein databases. J Mol Biol 251:390–399

Hobohm U, Houthaeve T, Sander C (1994) Amino acid analysis and protein database compositional search as a rapid and inexpensive method to identify proteins. Anal Biochem 222:202–209

James P, Quadroni M, Carafoli E, Gonnet G (1993) Protein identification by mass profile fingerprinting. Biochem Biophys Res Commun 195:58–64

Jensen ON, Podtelejnikov A, Mann M (1996) Delayed extraction improves specificity in database searches by MALDI peptide maps. Rapid Commun Mass Spectrom 10:1371–1378

Jonscher KR, Yates JR 3rd (1997) The quadrupole ion trap mass spectrometer — a small solution to a big challenge. Anal Biochem 244:1–15

Jungblut P, Dzionara M, Klose J, Wittman-Liebold B (1992) Identification of tissue proteins by amino acid analysis after purification by two-dimensional electrophoresis. J Prot Chem 11:603–612

Klose J (1975) Protein mapping by combined isoelectric focusing and electrophoresis in mouse tissues. A novel approach to testing for individual point mutations in mammals. Humangenetik 26:231–243

Latter GI, Burbeck S, Fleming J, Leavitt J (1984) Identification of polypeptides on two-dimensional electrophoresis gels by amino acid composition. Clin Chem 30:1925–1932

Liang X, Bai J, Liu YH, Lubman DM (1996) Characterisation of SDS-PAGE separated proteins by matrix-assisted laser desorption / ionisation mass spectrometry. Anal Chem 68:1012–1018

Link AJ (ed) (1997) 2-D gel electrophoresis protocols. Humana Press, in press

Maillet I, Lagniel G, Perrot M, Boucherie H, Labarre J (1996) Rapid identification of yeast proteins on two-dimensional gels. J Biol Chem 271:10263–10270

Mann M, Talbo G (1996) Developments in matrix-assisted laser desorption/ionisation peptide mass spectrometry. Curr Opin Biotechnol 7:11–19

Mann M, Wilm M (1994) Error tolerant identification of peptides in sequence databases by peptide sequence tags. Anal Chem 66:4390–4399

Mann M, Hojrup P, Roepstorff P (1993) Use of mass spectrometric molecular weight information to identify proteins in sequence databases. Biol Mass Spectrom 22:338–345

Neubauer G, Gottschalk A, Fabrizio P, Seraphin B, Lührmann R, Mann M (1997) Identification of the proteins of the yeast U1 small nuclear ribonucleoprotein complex by mass spectrometry. Proc Natl Acad Sci USA 94:385–390.

O'Farrell PH (1975) High resolution two-dimensional electrophoresis of proteins. J Biol Chem 250:4007–4021

Ou K, Wilkins MR, Yan JX, Gooley AA, Fung Y, Scheumack D, Williams KL (1996) Improved high-performance liquid chromatography of amino acids derivatised with 9-fluorenylmethyl chloroformate. J Chrom A 723:219–225

Packer NH, Wilkins MR, Golaz O, Lawson M, Gooley AA, Hochstrasser DF, Redmond JW, Williams KL (1996) Characterisation of human plasma glycoproteins separated by 2-D gel electrophoresis. Bio/Technology 14:66–70

Pappin DJC, Hojrup P, Bleasby AJ (1993) Rapid identification of proteins by peptide-mass fingerprinting. Curr Biol 3:327–332

Patterson DH, Tarr GE, Regnier FE, Martin SA (1995) C-terminal ladder sequencing via matrix-assisted laser desorption mass spectrometry coupled with carboxypeptidase Y time-dependent and concentration-dependent digestions. Anal Chem 67:3971–3978

Patterson SD, Thomas D, Bradshaw RA (1996) Application of combined mass spectrometry and partial amino acid sequence to the identification of gel-separated proteins. Electrophoresis 17:877–891

Rasmussen HH, Mortz E, Mann M, Roepstorff P, Celis JE (1994) Identification of transformation sensitive proteins recorded in human two-dimensional gel protein databases by mass-spectrometric peptide mapping alone and in combination with microsequencing. Electrophoresis 15:406–416

Roepstorff P (1997) Mass spectrometry in protein studies from genome to function. Curr Opin Biotechnol 8:6–13

Rougeon F, Kourlisky P, Mach B (1975) Insertion of rabbit beta-globin gene sequence into an *E. coli* plasmid. Nucl Acids Res 2:2365–2378

Saiki RK, Scharf S, Faloona F, Mullis KB, Horn GT, Erlich HA, Arnheim N (1985) Enzymatic amplification of beta-globin genomic sequences and restriction site analysis for diagnosis of sickle cell anemia. Science 230:1350–1354

Sanger F, Nicklen S, Coulson AR (1977) DNA sequencing with chain-terminating inhibitors. Proc Natl Acad Sci USA 74:5463–5467

Schena M (1996) Genome analysis with gene expression microarrays. Bioessays 18:427–431

Shaw G (1993) Rapid identification of proteins. Proc Natl Acad Sci USA 90:5138–5142

Shevchenko A, Jensen ON, Podtelejnikov AV, Sagliocco F, Wilm M, Vorm, O, Mortensen P, Shevchenko A, Boucherie H, Mann M (1996) Linking genome and proteome by mass spectrometry: large-scale identification of yeast proteins from two dimensional gels. Proc Natl Acad Sci USA 93:14440–14445

Spengler B, Kirsch D, Kaufman R, Jaeger E (1992) Peptide sequencing by matrix-assisted laser-desorption mass spectrometry. Rapid Commun Mass Spectrom 6:105–108

Sutton CW, Wheeler CH, U S, Corbett JM, Cottrell JS, Dunn MJ (1997) The analysis of myocardial proteins by infrared and ultraviolet laser desorption mass spectrometry. Electrophoresis 18:424–431

Thiede B, Wittmann-Liebold B, Bienert M, Krause E (1995) MALDI-MS for C-terminal sequence determination of peptides and proteins degraded by carboxypeptidase Y and P. FEBS Lett 357:65–69

Vestling MM, Fenselau C (1994) Poly(vinylidene difluoride) membranes as the interface between laser desorption mass spectrometry, gel electrophoresis, and in situ proteolysis. Anal Chem 66:471–477

Vorm O, Roepstorff P, Mann M (1994) Improved resolution and very high sensitivity in MALDI TOF of matrix surfaces made by fast evaporation. Anal Chem 66:3281–3287

Wang R, Chait B (1994) High-accuracy mass measurement as a tool for studying proteins. Curr Opin Biotechnol 5:77–84

Wasinger V, Cordwell SJ, Cerpa-Poljak A, Gooley AA, Wilkins MR, Duncan M, Williams KL, Humphery-Smith I (1995) Progress with gene-product mapping of the mollicutes: *Mycoplasma genitalium*. Electrophoresis 16:1090–1094

Wheeler CH, Berry SL, Wilkins MR, Corbett JM, Ou K, Gooley AA, Humphery-Smith I, Williams KL, Dunn MJ (1996) Characterisation of proteins from 2-D gels by matrix-assisted laser desorption mass spectrometry and amino acid compositional analysis. Electrophoresis 17:580–587

Wilkins MR, Williams KL (1997) Cross-species protein identification using amino acid composition, peptide mass fingerprinting, isoelectric point and molecular mass: a theoretical evaluation. J Theor Biol 186:7–15

Wilkins MR, Sanchez JC, Gooley AA, Appel RD, Humphery-Smith I, Hochstrasser DF, Williams, KL (1995) Progress with proteome projects: why all proteins expressed by a genome should be identified and how to do it. Biotechnol Gene Eng Rev 13:19–50

Wilkins MR, Pasquali C, Appel RD, Ou K, Golaz O, Sanchez JC, Yan JX, Gooley AA, Hughes G, Humphery-Smith I, Williams KL, Hochstrasser DF (1996a) From proteins to proteomes: Large scale protein identification by two-dimensional electrophoresis and amino acid analysis. Bio/Technology 14:61–65

Wilkins MR, Gasteiger E, Sanchez JC, Appel RD, Hochstrasser DF (1996b) Protein identification with sequence tags. Curr Biol 6:1543–1544

Wilkins MR, Ou K, Appel RD, Sanchez JC, Yan JX, Golaz O, Farnsworth V, Cartier P, Hochstrasser DF, Williams KL, Gooley AA (1996c) Rapid protein identification using N-terminal "sequence tag" and amino acid analysis. Biochem Biophys Res Commun 221:609–613

Wilkins MR, Gasteiger E, Tonella L, Ou K, Sanchez JC, Tyler M, Gooley AA, Appel RD, Williams KL, Hochstrasser DF (1997a) Protein identification with N- and C-terminal sequence tags in proteome projects. Submitted.

Wilkins MR, Lindskog I, Gasteiger E, Bairoch A, Sanchez JC, Hochstrasser DF, Appel RD (1997b) Detailed peptide characterisation using PEPTIDEMASS — a World-Wide Web accessible tool. Electrophoresis 18:403–408

Wilm M, Mann M (1996) Analytical properties of the nano electrospray ion source. Anal Chem 66:1–8

Wilm M, Shevchenko A, Houthaeve T, Breit S, Schweigerer L, Fotsis T, Mann M (1996) Femtomole sequencing of proteins from polyacrylamide gels by nano-electrospray mass spectrometry. Nature 379:466–469

Yan JX, Wilkins MR, Ou K, Gooley AA, Williams KL, Sanchez JC, Golaz O, Pasquali C, Hochstrasser DF (1996) Large scale amino acid analysis for proteome studies. J Chrom A 736:291–302

Yates JR III, Speicher S, Griffin PR, Hunkapiller T (1993) Peptide mass maps: a highly informative approach to protein identification. Anal Biochem 214:397–408

Yates JR III, Eng JK, McCormack AL (1995) Mining genomes: correlating tandem mass spectra of modified and unmodified peptides to sequences in nucleotide databases. Anal Chem 67:3202–3210

4 The Importance of Protein Co- and Post-Translational Modifications in Proteome Projects

Andrew A. Gooley and Nicolle H. Packer

4.1 Introduction

Any protein in a proteome can be modified by co- and post-translational modifications. There are literally dozens of different types of such modifications, all of which can influence a protein's charge, hydrophobicity, conformation and/or stability. Hence, the "one-gene-one-polypeptide" paradigm is now outdated as we understand that in both eukaryotes and prokaryotes the polypeptide translation of many genes is modified to create multiple gene products from a single DNA sequence (Fig. 4.1). Well-documented examples include the differential glycosylation of Thy-1 expressed in different tissues (Parekh et al. 1987) and the differential RNA splicing of N-CAM (Cunningham et al. 1987).

Specific amino acid sequences in a polypeptide tell a cell to process that polypeptide or add a certain post-translational modification. While many different types of modification have already been identified, most have not been localised on a sufficiently large number of proteins to allow the construction of extensive modification-specific databases that can predict all modifications from gene sequences. So the reverse arrow in Fig. 4.1, from protein modification to gene, continues to provide most of our knowledge on the structural and chemical features of the protein sequences which specify a particular type of modification. Table 4.1 lists some of the best studied examples of co- and post-translational modifications that are of known protein sequence specificity. It is not our intention to attempt a comprehensive review of protein co- and post-translational modifications here as the field has previously been extensively covered. There are several excellent reviews and book chapters which discuss many of the functions of co- and post-translational modifications as well as the methods used for their identification (Aitken 1990; Rucker and McGee 1993; Graves et al. 1994; Kellner et al. 1994). Here we aim to highlight how the proteome is influenced by changes that occur to the translated gene product in the different compartments that constitute the cell.

An important feature of proteome projects is that many of the protein isoforms generated by co- and post-translational modifications can be separated by two-

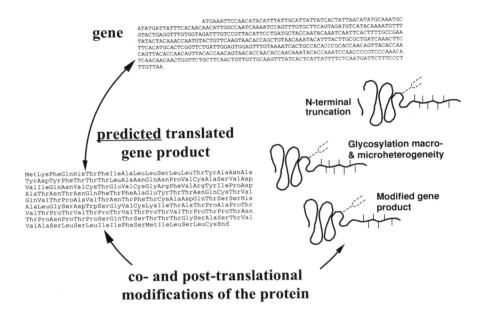

Fig. 4.1. Proteome diversity: from one gene sequence to one predicted protein sequence to many protein products. The protein described is the *Dictyostelium discoideum* cell surface protein PsA. When at the cell surface, it has lost its N-terminal signal peptide, has a C-terminal truncation with the addition of a GPI-anchor and is extensively O-glycosylated

dimensional polyacrylamide gel electrophoresis (2-D PAGE). This provides the research scientist with a powerful strategy to investigate how co- and post-translational modifications influence protein structure and function, and whether or not the expression of particular isoforms is under developmental or disease control. Following protein identification by any of the means described in Chap. 3, many of the modifications can not be reliably predicted from the DNA sequence, so analytical methods are needed not only to identify a modification but also to localise where it is on the protein.

Typically, the biochemical analysis of co- and post-translational modifications has involved either a chemical or enzymatic release of the modifications from a pool of protein isoforms. While the introduction of analytical 2-D PAGE provided a further dimension to the separation and characterisation of protein isoforms (Aebersold et al. 1987), rarely were each of the discrete protein entities characterised. As the technique of 2-D electrophoresis became increasingly popular following the commercialisation of immobilised pH gradient (IPG) strips (see Chap. 2), the most commonly described modification was the blockage of the protein α-amino group — an assumption often made when no initial yield of a phenylthiohydantoin amino acid was found in the first cycle of Edman protein sequencing (Bjel-

Table 4.1. Sequence motifs for some of the common co- and post-translational modifications. Amino acids in italics are those that will be modified in each case

Modification	Sequence	Comment
C-terminal GPI anchor	Ser or Gly or Cys—↓—Ala or Gly	While there is no defined consensus at the site of the signal sequence cleavage, these residues are commonly found[a,b,c]
Phosphorylation	Aaa–Aaa–X–Aaa–*Tyr*	Sites recognised by Tyr kinase
	X–*Ser/Thr*–Neg–Neg–Neg	Sites recognised by casein kinase II
	Lys/Arg–X–X–*Ser/Thr*–X– X–X–Lys/Arg	Sites recognised by protein kinase C[d]
	X–*Ser/Thr*–Pro–X or *Ser/Thr*–Pro–X	Sites recognised by p45 casein kinase[e]
Sulfation	Neg–Neg–*Tyr*–Neg–Neg	Tyr-SO$_4$ is frequently found at clusters of negatively charged amino acids
Glycosylation	*Asn*–X–Ser/Thr/Cys–X X≠P	N-linked glycosylation at Asn
	X–X–*Ser/Thr*–X–X X=Pro/Ser/Thr/Ala	O-linked glycosylation of GalNAc at Ser and Thr[f]
N-myristoylation	*Gly*–X–X–X–Ser/Thr	Myristate attached to Gly[g]
Hydroxylation	Gly–X–*Pro*–Gly–X	Hydroxylation of Pro is influenced by X
N-methylation	N–methyl*Met*–X–X–P–X	For methylation in *E. coli* see ref.[h]
Carboxymethylation	*Glu*–*Glu*–X–X–Aaa–Ser/Thr	For carboxymethylation at Glu see ref.[i]
Signal Peptidase site	Aaa–X–Baa–↓–X	Cleavage of the N-terminal signal peptide[j]
Prenylation	Gly–Cys–X–X–Cys–X–*Cys*– Aaa–Aaa–X–CO$_2$H	Prenylation at Cys in the Cys-Aaa-Aaa-X box. Typical of ras proteins[k]

Amino acids are represented with the standard three-letter code. Aaa is any aliphatic residue (Ile, Leu, Val and can be Ala), X is any amino acid, Neg is Asp or Glu, Baa is Ala, Gly or Ser. [a]Barclay et al. 1993; [b]Antony and Miller 1994; [c]Udenfriend and Kodukula 1995; [d]House et al. 1987; [e]Angelov 1994; [f]Hansen et al. 1995; [g]Johnson et al. 1994; [h]Apostol et al. 1995; [i]Terwilliger et al. 1986; [j]Perlman and Halvorson 1983; [k]Zhang and Casey 1996.

lqvist et al. 1994; Wirth et al. 1995; Posch et al. 1995). However, in many cases picomole amounts of protein are sequenced and it is difficult to determine whether there is insufficient protein or an authentic blocked N-terminus (see Sect. 4.4.1).

Many protein isoforms separated by 2-D PAGE appear as a "train" of protein spots differing in pI and/or apparent mass that is suggestive of glycosylation or phosphorylation. Other isoforms show no clear 2-D gel patterns, such as N- or C-terminal truncations of a protein. These and many other modifications can only be identified by the analysis of discrete protein spots.

Micropreparative 2-D PAGE (see Chap. 2) has heralded a new era in the analysis of co- and post-translational modifications, as increases in the loading capacity of the first dimension allow several micrograms of protein to be resolved into single spots on a single gel. Hence, there is sufficient quantity of protein in a single spot to allow several compositional analyses for the well characterised post-translational modifications such as phosphoamino acids, neutral, acidic and amino sugars and many modified amino acids (Packer et al. 1996; Yan et al. 1997; Grant and Crankshaw 1997). Traditional technologies for the analysis of co- and post-translational modifications such as the incorporation of a radiolabel, HPLC (for example high performance anion exchange chromatography of sugars) and the Edman sequenator (see Grant and Crankshaw 1997) remain practical and user friendly. However, the technology which appears to have the greatest potential for the analysis of co- and post-translational modifications is the mass spectrometer (MS) (Carr et al. 1991). This technology is gradually moving from specialist laboratories into protein core facilities, yet there are many traps which tend to be overlooked in the interpretation of the mass spectra. These include the close mass similarity between sulfate and phosphate groups and the fact that sugars which have quite different functions can have an identical mass (e.g. the N-acetylhexosamines). In the reporting of mass spectrometric identification of full oligosaccharide structures there is often an absence of the monosaccharide and exoglycosidase analyses needed to identify the specific sugars present in the structure. While it is certainly exciting to obtain an exact mass match to an expected structure, and the structures deduced solely from MS analysis can often be correct, protein chemists should realise that there are some limitations to the technique. In Sect. 4.4 we will discuss the role of MS in combination with other techniques in detecting and analysing co- and post-translational modifications.

4.2 An overview of modifications: what are they and where do they occur?

The potential diversity of the final protein product from any gene is staggering when it is considered that there are several hundred types of modifications which can alter the protein over time (such as in development and differentiation) and in space (such as in its location within the cell). An important consideration in proteome analysis is that unlike the genome, which can be seen as a static information repository, the proteome can be compartmentalised into different organelles or into

Table 4.2. Compartmentalisation of protein modifications by subcellular location

Compartment	Sub-compartment	Type of modification
Intracellular	Nucleus	Acetylation (histone acetylation at ε-amino), phosphorylation, O-GlcNAc
	Lysosome	Mannose-6-phosphate labelled N-linked sugars
	Mitochondria	N-formyl acylation
	Chloroplast	N-formyl acylation, pigments and light harvesting groups (e.g. chlorophyll)
	Golgi	N- and O-linked oligosaccharide, sulfation, palmitoylation
	Endoplasmic reticulum	N-linked oligosaccharide, GPI-anchor
	Cytosol	Acetylation, methylation, phosphorylation, O-GlcNAc
	Ribosome	Myristoylation
Cell surface	Plasma membrane	N- and O-glycosylation, GPI-anchor
Extracellular	Extracellular fluid	N- and O-glycosylation, acetylation, phosphorylation
	Extracellular matrix	Hydroxylation, phosphorylation, N- and O-glycosylation

essentially three categories of protein: intracellular, cell-surface associated and extracellular (Table 4.2).

Thus by separating the cell into its constituent compartments, we have in some cases defined the types of modification that can occur to any particular protein in the proteome. For example, the covalent attachment of asparagine-linked oligosaccharides should only be found on endoplasmic reticulum, golgi, cell-surface, lysosomal and extracellular proteins. Since many of the modifications are due to enzymatic reactions, there is an increased probability that proteins derived from the same compartment as a certain modifying enzyme will be modified by that enzyme. For example, acetylation of the histones occurs in two compartments: lysine residues are predominantly acetylated in the nucleus, while α-amino groups are acetylated in the cytosol (Lopez-Rodas et al. 1991; Yamada and Bradshaw 1991). In many cases, the modification arises in transit through a compartment: for example, most cell-surface proteins which transit through the golgi contain one or two N-glycosylation sites and several O-linked glycosylation sites.

4.3 Modifications that influence protein charge on 2-D PAGE

The advances in 2-D PAGE have been addressed in Chap. 2 and 3. The principal merits of this separation technique are the high reproducibility of the separation and the high loading capacity. The modification of proteins with phosphate, sulfate or carboxyl groups (e.g. the sugar sialic acid) are direct ways of introducing new negative charges onto the proteins which may result in multiple isoforms after separation by 2-D PAGE (Hughes et al. 1992). The phenomenon of "trains" of spots is readily visible on eukaryotic protein reference maps in the SWISS-2DPAGE database[1]. These trains of spots can vary with disease states and in different tissues. For example, the pattern of spots for serotransferrin from human sera and cerebrospinal fluid are quite different (Wilkins et al. 1996), as are the patterns of transferrin spots in normal patients and those with alcohol-related disease (Gravel et al. 1996). Table 4.3 lists some of the common modifications as well as the modified amino acids that can produce a charge-dependent change to a protein.

4.4 Analysis of co- and post-translational modifications after 2-D PAGE

While there are few reports on the effects that the modifications in Table 4.3 have on the 2-D separation of proteins, there are three co- and post-translational modifications that have been studied in some detail: N-acetylation, glycosylation and phosphorylation. Before we discuss these three modifications we will first address the electrophoretic methods for preparing and detecting modified proteins.

4.4.1 Presentation of the modified protein

Gel versus membrane. Essentially there are two ways of analysing proteins separated by 2-D PAGE: *in situ* in the acrylamide matrix or electroblotted onto a membrane (see Fig. 4.2). Electroblotting onto membranes has revolutionised the throughput of samples in the Edman sequenator (Matsudaira 1987) as has the simultaneous rapid purification of many proteins that can now be achieved by 2-D electrophoresis. Whether or not to electroblot for high throughput protease digestion of proteins in proteome studies is a frequently debated topic. There are advantages and disadvantages to both *in situ* and blotting approaches (see Table 4.4). In a

---

[1] http://www.expasy.ch/ch2d/ch2d-top.html

Table 4.3. Modifications which lead to a charge-dependent change to a protein

Modification	Amino acids affected and comments
Acylation	Loss of the α-amino positive charge. Predominantly the addition of an acetyl (typically to Ser or Ala) or a pyroglutamyl group (pyrilidone carboxylic acid modification of Gln). In addition formyl, pyruvoyl, α-ketobutyryl, glucuronyl, α-aminoacyl groups are found on the α-amino group. Other modifications include fatty acylation, myristoylation, palmitoylation and isoprenylation.
Alkylation	Alteration of the α- or ϵ-amino positive charge. The predominant alkyl group is a methyl, which can be mono-, di- or trimethyl. Well known examples are the methyl derivatives of lysine[a] in histones, and histidine in actin and the myosin light chain kinase. Other methylated amino acids include Arg, Phe and carboxyl residues (see below). Only monomethyl derivatives of the α-amino group are suitable for Edman degradation (the exceptions are N-methylMet/Ala/Phe/Pro which block the α-amino group).
Carboxylmethylation	Esterification of specific protein carboxyl groups by methyltransferases.
Phosphorylation	Predominantly modifications to Ser, Thr and Tyr, but less commonly also Cys and His. Direct linkage to these amino acids or indirect links via oligosaccharides increases the negative charge[b].
Sulfation	Predominantly on Tyr and on oligosaccharides attached to Asn, Thr and Ser (e.g. mucins in cystic fibrosis[c]).
Carboxylation	γ-carboxyglutamate and β-carboxyaspartate both have two neighbouring carboxyl groups and can have two negative charges but with different pK values.
Sialylation	Predominantly on oligosaccharides attached to Asn, Thr and Ser.
Proteolytic processing	Truncation of protein N- and C-terminal regions which contain charged amino acids will lead to an alteration in pI, while not necessarily resulting in a substantial shift in Mr.

[a]The methyl derivatives of Lys can adopt different charges depending on the pH. Both mono- and dimethyl-Lys are alkyl amines that can become protonated, while trimethyl-Lys is a quaternary amine which is always positively charged. [b]The phosphodiester linked oligosaccharides are labile to mild acid, so caution must be taken when handling this type of modification. [c]Lo-Guidice et al. (1994).

Table 4.4. Practical aspects of analysis of proteins separated by 2-D PAGE

PROTEINS REMAINING *in situ*	
Advantages	Disadvantages
Endoproteinase Digests	
Dehydration of acrylamide spot and rehydration with an endoproteinase is an established technique in many laboratories[a]. This process has been automated[b]. Most modified peptides are recovered although there is little data concerning the recovery of glycopeptides.	Multi-step procedure, where some peptides can potentially be lost. Incomplete cleavage, non-specific cleavage and enzyme autolysis appear with variable degrees of frequency[c]. High salt buffers can be a disadvantage in MS analysis.
Analysis of Intact Protein	
Protein spots can be passively eluted from the acrylamide. Eluate is easily concentrated on the Applied Biosystems Division (ABD) ProSorb or the HP biphasic columns for N-terminal analysis. ProSorb can be used for amino acid analysis.	Inefficient procedure, often with poor recovery. High salt and detergent buffers make C-terminal sequencing and amino acid analysis impractical. Not practical for samples present in low picomolar amounts.

PROTEIN SPOTS TRANSFERRED TO MEMBRANE	
Advantages	Disadvantages
Endoproteinase Digests	
Simple procedure where the enzyme and buffer are added to the protein spot[d]. Can be automated[e]. Peptide digest has potential to be analysed directly off the membrane. However, there is little data concerning the recovery of modified peptides, although specific groups, such as oligosaccharides, are recovered in high yield (see Sect. 4.4.3).	Depending on the membrane, there can be poor recovery of peptides. There is insufficient data to determine how frequently the specific proteinases cut (see Lui et al. 1996). In most cases the membrane must be blocked with detergent to prevent adsorption of the endoproteinase.
Analysis of Intact Protein	
Simple procedure where the membrane spot is excised then placed into the protein sequenator, or hydrolysed for either amino acid analysis, phosphoamino acid analysis, or monosaccharide analysis. Requires no preparation and the membrane sample can be used for amino acid analysis after N-terminal tag. Direct mass analysis possible using Er-Yag-Laser (infrared wavelength) MALDI-MS[f].	While PVDF is a generic membrane for protein analysis, some membrane types can not be used for some applications. For example nitrocellulose is dissolved in the protein sequenator or charred during protein hydrolysis for amino acid analysis. Only Teflon™, Zitex™ and GoreTex™ can be used in the C-terminal sequenator as other membranes, including PVDF, are destroyed by the chemistry.

[a]Hellman et al. 1995; [b]Houthaeve et al. 1995; [c]Clauser et al. 1995; [d]Pappin et al. 1995; [e]Ducret et al. 1996; [f]Eckerskorn et al. 1997.

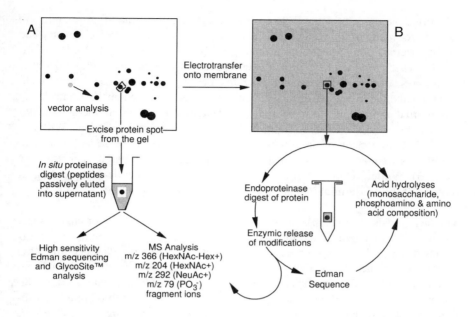

Fig. 4.2. The strategies used for the analysis of modified proteins separated by 2-D PAGE. (A) shows the techniques used for proteins remaining in the acrylamide matrix and (B) the strategies used for electroblotted samples. Following endoproteinase digests, or an enzymatic release of the modifications from proteins electroblotted onto PVDF, the products can be analysed by any of the procedures shown for the *in situ* analysis of proteins. GlycoSite is an Edman degradation procedure which identifies glycoamino acids (Gooley and Williams 1997). Note that the strategies described here can supply information for the identification of proteins as well as analysis of their post-translational modifications. For an example of vector analysis see Fig. 8.4, p. 147

recent study by the Association of Biomolecular Resource Facilities, which compared the success of both strategies from four high-profile US core facilities (the report is available on the Internet[2]), no particular method was favoured, although there is a general trend towards the *in situ* procedure based on the assumption that electroblotting onto membranes is an inefficient process. For reasons outlined below, we favour electroblotting to PVDF membranes for the analysis of glycoproteins and phosphoproteins (Table 4.4). A detailed study by Lui et al. (1996) has established the optimal conditions for the digestion and release of peptides from proteins electroblotted onto nitrocellulose.

---

[2] http://www.medstv.unimelb.edu.au/ABRFNews/1993/December1993/
dec93methodsforum.html

In principle the peptides recovered from the digestion of modified proteins are suitable for many of the well established techniques such as Edman chemistry and liquid chromatography-mass spectrometry (LC-MS), although only a few examples have been reported from proteins separated by 2-D PAGE (see below). Despite the progress of the LC-MS technology, Edman chemistry remains a powerful technique for the identification of modified amino acids and an extensive compilation of modified amino acids from a variety of phenylthiohydantoin-amino acid analysis systems is now available (Grant and Crankshaw 1997; Gooley and Williams 1997). The processing of protein N-termini, a further post-translational modification that is frequently ignored, is also easily detected by Edman sequencing (see Chap. 3). While protein C-termini can be processed, the chemical sequencing of proteins at their C-termini remains challenging (Bailey et al. 1995).

Vector analysis. The identification of modified proteins using vector analysis is a direct *in situ* approach. If an unfolded protein is to migrate to its predicted pI, it is assumed that it does so as a random coil so that only the amino acid composition contributes to the pI and there is no influence from other parts of the protein (Gianazza 1995). When proteins are separated by 2-D PAGE under denaturing conditions (e.g. dissolving the sample in urea and thiourea with dithiothreitol or tributyl phosphine), the migration position of unmodified proteins on 2-D PAGE typically corresponds to the position predicted from the amino acid sequence (Bjellqvist et al. 1993, 1994). Assuming that every precaution is taken with sample preparation (e.g. reduced thiols are alkylated with iodoacetamide or acrylamide monomer), the migration of a known protein to a position on the 2-D gel which is different from that predicted is strongly suggestive of a co- or post-translational modification.

Wilkins et al. (1996) have taken this observation to a more sophisticated level. They represent the difference between the observed and expected migration position of a protein with a line called a vector, and have created "vector maps" of the well characterised 2-D reference maps (see Fig. 8.4, p. 147). One example they showed is a vector map constructed for a region of the *E. coli* 2-D reference map. While many proteins showed no vector, those that had shifted in the gel did so almost exclusively in a monodimensional manner, showing a change in only one of the pI or MW axes. Much more complexity is observed with vector analysis of eukaryotic proteomes, where typically a single protein is seen as a string of different post-translationally modified forms (Wilkins et al. 1996).

4.4.2 Detection of co- and post-translational modifications

The initial step, once a 2-D separation has been achieved, is to identify which spots carry modifications of interest. Differential, sensitive staining of proteome maps for specific co- and post-translational modifications needs to be developed as an adjunct to the general protein stains for comparative mapping purposes. Ideally,

Table 4.5. Methods used for detection of post-translational modifications on gels or blots

Detection Method	Medium	Sensitivity[h]	Specificity
Monoclonal antibodies[a]	nitrocellulose, PVDF	10 ng	specific epitopes e.g. sugar, phosphate
Metabolic labelling[b]	SDS gel, PVDF, nitrocellulose	50 ng	specific precursors e.g. ^{32}P, ^{3}H-GalNAc
Lectins[c]	nitrocellulose, PVDF	0.1 µg	may be specific to one monosaccharide
Digoxenin[c,d]	nitrocellulose, PVDF	0.1 µg	vicinal hydroxyl groups of sugars
PAS Stain[c,e,f]	SDS gel, PVDF, nitrocellulose	1-10 µg	vicinal hydroxyl groups of sugars
Monosaccharide analysis[g]	PVDF	5 µg	all monosaccharides

[a]Wang 1988; [b]Hildebrandt and Fried 1989; [c]Packer et al. 1997; [d]Boehringer-Mannheim DIG detection kit; [e]Devine and Warren 1990; [f]Thornton et al. 1996; [g]Packer et al. 1996; [h]The sensitivity will depend on extent of modification of the protein.

sequential specific stains should be the first step in identifying alterations induced by disease, development and other cellular processes. For example, a single blot could be subjected to different stains to pinpoint glycosylation and phosphorylation, both of which may occur on the same protein. Preferably the sensitivity required for proteome analysis would be obtained by fluorescence rather than radiolabel. It would be useful for fluorescent adducts to have different excitation and/or emission maxima to allow discrimination between the different modifications.

Various methods have been developed for the specific detection of modified proteins on 2-D gels and blots, and these can be carried out at the analytical level. Most of these methods are destructive and, apart from specific lectin reactivity of glycoproteins, give little structural information. They are usually used as an identification tool for targeting those proteins of interest for further investigation. Some examples of methods which we have found to be the most useful for visualising, both on gels and blots, the phosphoproteins and glycoproteins separated by 2-D electrophoresis are shown in Table 4.5. The level of sensitivity of any method will depend on the extent of modification of the protein and the mechanism of detection. A note of caution should be raised regarding the transfer efficiency of modified proteins from gels to membranes, as highly glycosylated and/or charged proteins often do not electroblot efficiently and the standard general protein stains (e.g. Coomassie Blue) often do not visualise glycoproteins well.

4.4.3 Analysis of co- and post-translational modifications

Since many of the modified proteins are present in low abundance, the analytical procedures used must be highly sensitive. Ideally, several analyses should be able to be carried out sequentially on the same 2-D protein spot to maximise the data obtained and allow for automation of the analyses.

The characterisation of protein co- and post-translational modifications from gels and blots after 2-D PAGE separation can be approached in two ways: where modifications are still attached to the protein or after their release. Techniques available for protein sequence analysis have been fine-tuned to allow detailed characterisation of modifications at the low picomole level. Both Edman sequencing and mass spectrometric methods can be applied. The co- and post-translational modifications that affect the primary amino acid sequence can thus often become apparent by obtaining unclear sequences of a known protein. In this manner, protein/peptide sequencing can identify N- and C-terminal trimming, internal deletions, phosphoamino acids (Meyer 1994; Aebersold et al. 1991) and site-specific glycosylation (Gooley and Williams 1997). Similarly, unidentified masses in a tryptic digest of a known protein can imply potential sites of acetylation, methylation, glycosylation, phosphorylation, sulfation, or other modifications (e.g. Ducret et al. 1996). Below we briefly discuss acetylation, phosphorylation and glycosylation in proteomes, and how these modifications can be analysed.

Acetylation. N-acetylation is the major eukaryotic co-translational modification of protein α-amino groups. One consequence of N-acetylation is that the electron withdrawing capacity of the carbonyl group, which is adjacent to the protein α-amino group, effectively forms an amide bond and thus blocks the α-amino group to the Edman degradation. Is this loss of a single positive charge enough to be detected by 2-D PAGE?

In order to determine how many proteins were substrates for an N-terminal acetyltransferase (N^{α}-acetyltransferase) in *Saccharomyces cerevisiae*, a mutant was isolated that lacked this activity (Mullen et al. 1989). 2-D PAGE was used to compare the soluble proteins from the wild type yeast with those from mutant cells that lacked N^{α}-acetyltransferase (Lee et al. 1989). Among 855 proteins found by 2-D PAGE in wild type and mutant yeast cells, 20% of the proteins either disappeared or migrated to a higher pI. Proteins at a higher pI did not change in apparent mass but shifted uniformly some +0.1 to +0.2 pI units, a shift consistent with the protonation of the α-NH$_2$ group. Interestingly, another 12% of proteins were observed to have either increased or decreased expression levels, suggesting that acetylation is involved in a regulatory role with protein expression (Lee et al. 1989). While only 20% of proteins were altered due to the deletion of the N^{α}-acetyltransferase gene, others have suggested figures much higher ($\geq$50%) for soluble N-acetylated proteins (Driessen et al. 1985). The presence of a family of N^{α}-acetyltransferases may account for these differences (Lee et al. 1989).

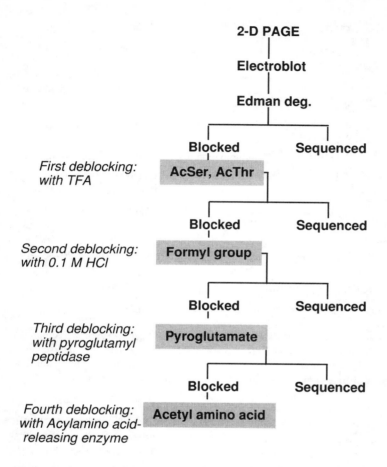

2-D PAGE

Electroblot

Edman deg.

	Blocked	**Sequenced**
First deblocking: with TFA	AcSer, AcThr	
	Blocked	**Sequenced**
Second deblocking: with 0.1 M HCl	Formyl group	
	Blocked	**Sequenced**
Third deblocking: with pyroglutamyl peptidase	Pyroglutamate	
	Blocked	**Sequenced**
Fourth deblocking: with Acylamino acid-releasing enzyme	Acetyl amino acid	

Fig. 4.3. Strategy for the sequential deblocking of N-terminally blocked proteins. Adapted from Hirano et al. (1993). Note that the fourth deblocking method works only on peptides, so an endoproteinase digestion is required before this step. While all acetyl amino acids can be liberated with an Acylamino acid-releasing enzyme, if a TFA deblocking step has already been used the acetyl Ser and acetyl Thr will have previously been analysed

Direct chemical approaches for the analysis of proteins N-terminally blocked by acetylation or other modifications are possible but they often lead to poor yields of the unblocked N-terminus as well as an increased likelihood of internal peptide bond cleavage (Hirano et al. 1993). Nevertheless, a sequential strategy of deblocking is possible on abundant proteins that have been transferred to PVDF (Fig. 4.3).

Phosphorylation. Traditionally, phosphorylated proteins have been found either by radiolabelling, which cannot be applied in many situations, or by phosphoamino

Table 4.6. Examples of glycoprotein and sugar analysis after SDS-PAGE

protein	separation medium	release method	detection	amount
asialo-α1-acid glycoprotein[a]	1-D PVDF, dried gels	gas-phase hydrazinolysis	radiolabel	3.5–28.5 µg
glycophorin A epitectin[b]	1-D nylon, NC, PVDF	β-elimination	radiolabel	50 µg
rG-CSF[c]	1-D PVDF	β-elimination	PA[i] fluorescence	30 µg
α-galactosidase A[d]	1-D PVDF	PNGase F	ANTS[j] fluorescence	20 µg
fetuin α1-antitrypsin αHS-glycoprotein[e]	1-D / 2-D PVDF	acid hydrolysis / PNGase	HPAEC-PAD[k]	2.5–20 µg
H+K+ATPase[f]	1-D PVDF	acid hydrolysis PNGase	HPAEC-PAD	20–30 µg
rG-CSF[g]	1-D IEF / PVDF	β-elimination	HPAEC-PAD	20 µg
interferon gamma[h]	1-D SDS gel	in-gel protease	MALDI-MS	2 µg

[a]Kawashima et al. 1992; [b]Zhu and Bhavanandan 1995; [c]Oh-eda et al. 1996; [d]Friedman and Higgins 1995; [e]Packer et al. 1996 and Fig. 4.5; [f]Weitzhandler et al. 1993; [g]Andersen et al. 1994; [h]Mörtz et al. 1996; [i]pyridylamino; [j]amino naphthalene trisulfonic acid; [k]High performance anion exchange chromatography with pulsed amperometric detection.

acid-specific antibodies on blots from 1-D and 2-D gels, which can be non-specific and expensive (Wang 1988; Goldstein et al. 1995). A method to find specific phosphoamino acids present on proteins separated by 2-D electrophoresis has been recently described which should allow for the high throughput analysis required for proteome mapping (Yan et al. 1997). Site identification of phosphorylation should be possible on 2-D separated proteins by Edman sequencing of thiol derivatised phosphopeptides obtained by protease digestion (Meyer 1994), and this approach is currently more effective than MS/MS sequencing of phosphopeptides which is limited to the identification of P-Tyr since collision-induced dissociation (CID) MS results in the loss of phosphate from P-Thr and P-Ser. However, mass spectrometry can offer a sensitive approach for detecting the specific mass addition of a phosphate to a peptide, using peptide mass fingerprinting of protease digests both with and without alkaline phosphatase treatment to check for loss of the specific mass of the phosphate group (Liao et al. 1994). Phosphopeptides have also been analysed by matrix-assisted laser desorption / ionisation (MALDI) (Zhang et al. 1994) or electrospray ionisation (ESI) mass spectrometry after 2-D thin layer chromatography mapping of phosphopeptides (Watts et al. 1995). A caveat here is, however, that the addition of either sulfate or phosphate results in approximately the same mass change (80 Da) to the peptide. Other techniques such as ion analysis may also be needed to identify the modification.

Glycosylation. The characterisation of protein glycosylation is extremely complex. This is because microheterogeneity can occur through the presence of different glycoforms at a single amino acid site, as can macroheterogeneity which results from the variable presence of glycoforms at different amino acids in different copies of a protein. It has been calculated that recombinant tissue plasminogen activator, which has 3 sites of completely characterised N-glycosylation, has potentially 11,520 possible isoforms (Appfel et al. 1995). Fortunately, biology seems to be relatively well-ordered so that glycoproteins, although they have a huge potential heterogeneity, do in fact separate by 2-D PAGE into only a small number of discrete isoforms. Why this occurs is still to be determined, but the fact that it does emphasises the importance of characterising the differences between these glycoforms. This characterisation is multidimensional if a detailed structure is required, and involves extensive analyses to determine a) the monosaccharide composition, b) the attachment site(s) to the protein, c) the sequence, branching and linkage positions and d) the anomeric configuration of the monosaccharides.

Glycoprotein analysis began in 1938 when Albert Neuberger used 120 g of ovalbumin to prove that the carbohydrate present was not a contaminant but was attached to the protein (Neuberger 1988). Today, most analyses of glycoproteins are carried out on recombinant products, presumably because of their commercial importance and their availability in relatively large (mg to g) quantities. Techniques and strategies have been developed to successfully characterise the glycosylation on glycopeptides from recombinant proteins, predominantly using mass spectrometric techniques such as CID ESI mass spectrometry (Roberts et al. 1995) and MALDI-MS with exoglycosidase treatment (James et al. 1995). These approaches have been extensively reviewed elsewhere (James 1996; Andersen et al. 1996). The chemically or enzymatically released oligosaccharides have been analysed by chromatography, capillary electrophoresis, mass spectrometry and NMR (reviewed by Hardy[3]). Highly specialised mass spectrometric techniques have been required to obtain and interpret the monosaccharide sequence of oligosaccharides, such as high energy CID mass spectrometry (Medzihradszky et al. 1996) or MALDI coupled to a magnetic sector mass spectrometer (Harvey et al. 1994).

Although strategies to approach the analysis of glycosylated proteins in gels and blots have been published (Packer et al. 1996; Townsend and Trimble 1996[4]), relatively little data has been published on the characterisation of glycoproteins which have been separated by 1-D or 2-D PAGE (Table 4.6). Presumably this is due to the

---

[3] http://www.medstv.unimelb.edu.au/ABRFNews/1994/April1994/apr94methods.html

[4] http://www.medstv.unimelb.edu.au/ABRFNews/1996/September1996/sep96glycoprot.html

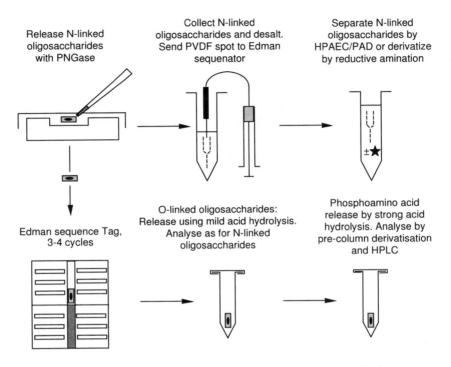

Fig. 4.4. Strategy for the sequential analysis of glycans, protein N-terminal sequence, and phosphoamino acids from a single protein isoform separated by 2-D PAGE

limited amount of material presented by this approach and the low sensitivity of many available sugar analysis methodologies which have been developed on large quantities of polysaccharides.

There are a limited but growing number of methods now available for high sensitivity analysis of the oligosaccharides of proteins.Edman sequencing of modified amino acids has enabled site identification and characterisation of glycopeptides at levels of sensitivity equivalent to that of protein analysis (Gooley and Williams 1997). Fluorescent labelling of released oligosaccharides by adducts such as ANTS and aminobenzamide allow detection of oligosaccharides at the picomole level. Mass spectrometry is now providing valuable glycosylation data on sub-picomole amounts of proteins and peptides, particularly where the appropriate exoglycosidases are available. The ease of characterisation of glycoforms has been improved by the increased amount of material now available for analysis from a 2-D PAGE blot. High-loading narrow-pH range IPG gels are available, as are methods for increasing protein solubility (see Chap. 2). Together these provide reproducible separations, which when combined with sample prefractionation based on mass, charge or affinity, can yield enough material for standard sugar analysis (Packer et al. 1996). Releasing the oligosaccharides enzymatically or chemically for analysis

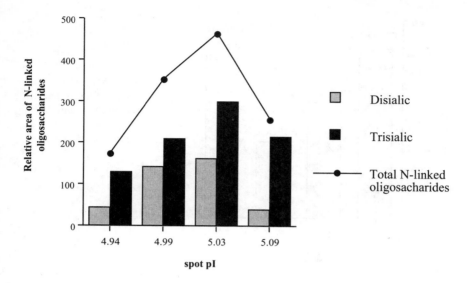

Fig. 4.5. Analysis of human plasma α1-antitrypsin glycoforms separated by 2-D PAGE and electroblotted onto PVDF. The N-linked oligosaccharides were released by PNGase and separated by high performance anion exchange chromatography. The pI values were estimated by 2-D PAGE

results in the loss of site-specific information but can help in determining the differences between classes of isoforms separated on 2-D gels. Clearly, it is desirable to get as much information as possible from each 2-D spot to verify each protein's identity and check the possibility that other co- and post-translational modifications are not responsible for the separation of isoforms.

We have achieved some progress towards the above mentioned goals in the sequential analysis of the plasma proteins, α1-antitrypsin and α-HS-glycoprotein separated by 2-D PAGE of human plasma (Packer et al. 1996). The sialic acid, neutral and amino sugar composition of the isoforms of these proteins, when electroblotted to PVDF, were determined prior to amino acid analysis and identification of the protein. More recently, we have been able to analyse the isoforms of these proteins by the strategy shown in Fig. 4.4.

The N-linked oligosaccharides of four isoforms of human plasma α1-antitrypsin (A1AT) separated by preparative 2-D PAGE and electroblotted to PVDF were released by PNGase F. The profiling of sialylated oligosaccharides released by the PNGase digestion of A1AT isoforms (Fig. 4.5) makes it clear that there is a complex mixture of modifications contributing to the 2-D PAGE separation of isoforms. Whereas an increased N-oligosaccharide branching and sialylation per mole of protein may account for the separation of the three most acidic spots of A1AT, it does not account for the migration of the most basic spot (pI 5.09). Thus simplistic

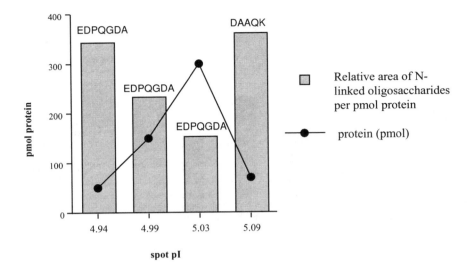

Fig. 4.6. N-terminal sequence analysis of human plasma α1-antitrypsin glycoforms following the release of the N-linked oligosaccharides by PNGase. The relative amount of N-oligosaccharides are expressed as units per pmol of protein, which was estimated from the N-terminal sequence tag (the tag is shown on top of each column). The pI values were estimated by 2-D PAGE

explanations or algorithms to explain migration differences between the isoforms may be difficult to formulate.

Following PNGase digestion, each of the PVDF spots — which still carried the protein sample — were subjected to N-terminal Edman degradation. This produced a sequence tag of 4 or more amino acids (see Sect. 3.3.4), and provided data concerning protein quantity and purity. Whilst the most basic spot had a similar oligosaccharide profile to the most acidic isoform (pI 4.94), the Edman sequence tag revealed that the isoform of pI 5.09 lacked the N-terminal pentapeptide (EDPQG) with two acidic amino acids, thus conferring a change to a more basic pI (Fig. 4.6).

The complexity of the effect of modifications on proteins, particularly on their migration pattern in isoelectric focusing, is illustrated in experiments whereby sialic acid is removed by neuraminidase treatment of protein isoforms (Packer et al. 1997; Hidaka and Fidge 1992). The "train" of spots on the 2-D separation moves to a more basic pI, but multiple spots covering a wide pI range are still often seen. Whilst this suggests that the remaining heterogeneity is the result of other charged modifications, one must interpret the products of a neuraminidase digestion with caution in the light of the different linkage specificities of the sialidase used, and the possibility that complete desialylation has not occurred. The analysis of 9 isoforms of serum amyloid A protein by Ducret et al. (1996) is a good example of the heterogeneity of modifications that can affect a single gene product. The amyloid

A protein exists in forms in which the N-terminal arginine is missing, some of the protein is glycosylated, there is oxidation at tryptophan residues and there is suspected modification in the form of dimethyl asparagine.

In many applications of proteome analysis, the fine structure of the oligosaccharides responsible for the separation of isoforms by 2-D PAGE may not need to be determined. However if they are to be characterised, it is probably necessary to use the approach of releasing the sugars from the separated protein spots in order to analyse the detailed differences such as linkage and anomeric configuration, particularly if specific exoglycosidases are unavailable. This will require improvements in the detection, separation and mass spectrometric analysis of the released oligosaccharides. Simple, non-destructive derivatisation chemistries which are nonselective are still required to make sugar molecules easier to separate, quantitate and sequence at high sensitivity. The aminobenzamide / anthranilic acid derivatives developed by Oxford GlycoSciences (Bigge et al. 1995) come close to achieving this goal, in that the derivatives give the oligosaccharides fluorescence and other useful electrophoretic, liquid chromatographic, and mass spectrometric properties.

4.4.4 Mass spectrometry perspectives in the analysis of protein co- and post-translational modifications

Mass spectrometry has exploded onto the biological analysis scene in recent years and is proving to be a powerful tool for the analysis of protein co- and post-translational modifications. Its role in proteome analysis will be extremely important, but the point should be made that mass spectrometry remains an adjunct to other analytical approaches rather than being an analytical "holy grail". This is because the data a mass spectrometer provides is the mass of a compound, or in the case of most analyses for post-translational modifications, the difference between the mass of a modified peptide and the same peptide without modifications. As a case in point, if the 5 different trisialylated oligosaccharides of fetuin are collected and submitted to mass spectrometry, only a single mass is seen, since the trisialics are isomeric structures of the same monosaccharide constituents (data not shown). In addition it must be noted that the various hexoses (glucose, mannose, galactose) and inositol all have the same monoisotopic residue mass (162.058 Da), as do glucosamine and galactosamine (161.069 Da), and there is no mass difference between the acetylated amino sugars GalNAc and GlcNAc (203.079 Da). The modifications of phosphate or sulfate have very similar monoisotopic residue masses (phosphate 79.663 Da, sulfate 79.957 Da), but CID fragmentation can be used to differentiate between them. Compositional analyses by other techniques are thus needed in these cases to corroborate the assignments. Having said that, the power of the mass spectrometer lies in its ability to confirm and suggest the structure of a co- or post-translational modification. In particular, if the sequence of a protein is known, the information that can be gained from mass spectrometry about modifications is considerable.

Table 4.7. Application of MALDI-TOF[a] MS and ESI-MS[b] to glyco/phosphoprotein analysis

ADVANTAGES	DISADVANTAGES
Intact Glyco/Phosphoproteins *MALDI-TOF*	
-high molecular mass range (up to 300 kDa) -molecular ions have low charge state which simplifies interpretation of heterogeneous modifications -tolerant to low concentrations of salts and detergents -mass accuracy has improved with delayed extraction technique ($\pm$ 0.05%) -high sensitivity by TOF detection	-molecular mass range less than 100 kDa with IR laser off blots -gives an indication of the mass range of modification but difficult to get detailed analysis of glycoforms -broad peaks obtained with heterogeneous proteins -variable crystallisation of proteins in the matrix results in non-ionisation of some protein samples particularly if in mixtures
ESI-MS	
-high mass accuracy ($\pm$ 0.01%) and resolution can resolve individual glycoforms -can be coupled to chromatography which is useful to separate the heterogeneous mixtures of glycoforms and to remove contaminants which may suppress the ionisation	-lower mass range (100 kDa) than MALDI -intolerant of salts and detergents -multiple charge states can make deconvolution and interpretation difficult, particularly for highly glycosylated proteins
Glyco/Phosphopeptides *MALDI-TOF*	
-only a fraction of a sample is consumed in an experiment as laser only irradiates a small area ($\sim$0.02 mm^2), allowing for multiple analyses on the one sample -allows sequencing of oligosaccharides on a glycopeptide with sequential exoglycosidases (Mörtz et al. 1996)	-usually need LC to separate modified peptides before applying individually to metal stage with matrix -improved matrices need to be developed for modified peptides -sialylation of glycopeptides decreases efficiency of ionisation
ESI-MS	
-in source CID and MS/MS enables identification of modified peptides using specific fragment ions -development of nanospray source enables multiple MS/MS sequencing experiments on the same sample -MS/MS can allow modified site determination	-acidic buffers used with CE interface may de-sialylate oligosaccharides
Released Oligosaccharides *MALDI-TOF*	
-can analyse neutral oligosaccharides without derivatisation -use of new matrix mixtures (e.g. Küster et al. 1996) has increased sensitivity and resolution	-acidic, sialylated oligosaccharides difficult to analyse -500 Da lower mass limit due to matrix interference restricts complete exoglycosidase digestion experiments and analysis of O-linked oligosaccharides
ESI-MS	
-desalting of isoforms of oligosaccharides separated by HPAEC allows analysis of individual oligosaccharides	-low sensitivity (pmol) of underivatised oligosaccharides -requires derivatisation of oligosaccharides for more sensitive detection

[a]Collision cells are available on the latest MALDI-TOF instruments.

[b]Note that ESI-TOF instruments may have advantages of both systems.

A recent review (James 1996) has compared the different modes of ionisation and detection used in mass spectrometry for the analysis of recombinant glycoproteins; these include FAB (fast atom bombardment), LSI (liquid secondary ion), MALDI-TOF and ESI-MS. The latter two ionisation techniques are proving to be the most useful for the analysis of co- and post-translational modifications, with each having advantages and disadvantages related to their method of ionising the molecules as well as in their respective use of time-of-flight (TOF) or quadrupole detection (see Table 4.7, Fig. 3.5 and 3.6). Interestingly, it is apparent that neither method is proving to be ideal for all applications. Further technology is emerging and a combination of electrospray ionisation coupled with ion-trap detection (ESI-IT) or with the increased sensitivity of time-of-flight detection (ESI-TOF) will further advance the mass spectrometry technique. However, it should be noted that the techniques which have been developed for non-modified proteins and peptides often prove to be inadequate for analysing the heterogeneity present in many post-translational modifications. As a rule, if there are a series of glycosylation or other modifications on a single protein or peptide, other complementary analyses will be needed to determine the structure of the modifications.

For proteome analysis, an accurate measurement of the true mass of a separated protein by MS, including its co- and post-translational modifications, is an invaluable attribute for identification purposes (see Sect. 3.3.3). However, it is well known that the apparent mass of, for example, glycoproteins estimated by PAGE can grossly overestimate the true mass (Gooley et al. 1997). The development of the IR laser along with different membranes and matrices for MALDI-TOF analysis of whole proteins blotted to Teflon™ membranes will hopefully provide this total mass data (Eckerskorn et al. 1997; Sutton et al. 1997), although currently this technology is limited to proteins less than 100 kDa.

Modified peptides released by in-gel or on-blot digestions have largely been ignored in the quest for protein identification by peptide mass fingerprinting. Since only a few "correct" peptide masses are required for identification of the protein from the database the remainder are usually not further analysed. These "incorrect" masses might provide information on protein co- and post-translational modifications once conditions for efficient recovery have been determined. However, improved software is needed to integrate the data from the mass spectrometer with possible mass additions due to co- and post-translational modifications. The predicted possible modifications could be checked on the targeted peptide by tandem MS peptide sequencing. Additional parameters could also be obtained for peptides by source CID-MS, to produce the diagnostic fragment ions of m/z 204 ($HexNAc^+$) and m/z 79 (PO_3). This scenario, of course, depends on the accuracy of the DNA databases as well as a high mass accuracy of the mass spectrometer and a comprehensive database of modifications likely to be encountered.

The technique of mass spectrometry has become central to the high sensitivity analysis of glycoproteins and other co- and post-translational modifications. Future strategies will combine this very sensitive approach with other appropriate techniques and integrate it with the necessary bioinformatics and automation required for proteome analysis.

4.5 Future directions

This chapter has outlined the important feature that distinguishes genome and proteome studies. The impact that co- and post-translational modifications have on final gene products is evident on the 2-D gel, where a diversity not encoded at the gene level is seen. We have also touched on some of the technologies used to analyse these modified proteins. While it is becoming clear that these co- and post-translational modifications are closely regulated and change depending on the tissue, developmental stage, disease state, or age of the cell, the importance of being able to identify, locate and characterise the differences induced by these modifications is paramount. 2-D PAGE remains the best way to display these changes, and this has resulted in an increasing focus on developing more sensitive technologies for analysing modified proteins. The technology is developing rapidly but is only just beginning to be applied to the protein isoforms separated on gels and/or transferred to membranes. With enough experimental data, vector analysis methods as outlined by Wilkins et al. (1996) may be able to predict and target the position(s) of the modified protein(s). The challenge remains to move beyond protein identification and to characterise the type and position of the modification at the low picomole level.

Acknowledgments

This chapter contains some original unpublished data (Sect. 4.4.3). We are grateful to Jean-Charles Sanchez for providing the PVDF blots of plasma proteins and Margaret Lawson for expert technical assistance. The authors acknowledge the support of the Australian Proteome Analysis Facility, the Australian Medical Research Council, and Australia Research Council.

References

Aebersold RH, Leavitt J, Saavedra RA, Hood LE, Kent SB (1987) Internal amino acid sequence analysis of proteins separated by one- or two-dimensional gel electrophoresis after in situ protease digestion on nitrocellulose. Proc Natl Acad Sci USA 84:6970–6974

Aebersold R, Watts JD, Morrison HD, Bures EJ (1991) Determination of the site of tyrosine phosphorylation at the low picomole level by automated solid-phase sequence analysis. Anal Biochem 199:51–60

Aitken A (1990) Identification of protein consensus sequences: active site motifs, phosphorylation, and other post-translational modifications. Ellis Horwood

Andersen DC, Goochee CF, Cooper G, Weitzhandler M (1994) Monosaccharide and oligosaccharide analysis of isoelectric focusing-separated and blotted granulocyte colony-stimulating factor glycoforms using high-pH anion-exchange chromatography with pulsed amperometric detection. Glycobiology 4:459–467

Andersen JS, Svensson B, Roepstorff P (1996) Electrospray ionization and matrix assisted laser desorption/ionization mass spectrometry: powerful analytical tools in recombinant protein chemistry. Nature Biotechnol 14:449–457

Angelov I (1994) Characterisation of a proline-directed casein kinase from bovine brain. Arch Biochem Biophys 310:97–107

Antony AC, Miller ME (1994) Statistical prediction of the locus of endoproteolytic cleavage of the nascent polypeptide in glycosylphosphatidylinositol-anchored proteins. Biochem J 298:9–16

Apostol I, Aitken J, Levine J, Lippincott J, Davidson, JS, Abbott-Brown D (1995) Recombinant protein sequences can trigger methylation of N-terminal amino acids in *Escherichia coli*. Prot Sci 4:2616–2618

Appfel A, Chakel J, Udiavar S, Hancock W, Souders C, Pungor E Jr (1995) Use of hyphenated liquid-phase analyses and mass spectrometric approaches for the characterisation of glycoproteins derived from recombinant DNA. In: Snyder AP (ed) Biochemical and biotechnological applications of electrospray ionization mass spectrometry. American Chemistry Society, Washington, pp 432–471

Bailey JM, Tu O, Issai G, Shively JE (1995) Automated C-terminal sequencing of polypeptides containing C-terminal proline. In: Atassi MZ and Appella E (eds) Methods in protein structure analysis. Plenum Press, New York, pp 131–138

Barclay AN, Birkeland ML, Brown MH, Beyers AD, Davis SJ, Somoza C, Williams AF (1993) The Leucocyte Antigen Facts Book. Academic Press, Harcourt Brace Jovanovich, London

Bigge JC, Patel TP, Bruce PJA, Goulding PN, Charles SM, Parekh RB (1995) Nonselective and efficient fluorescent labeling of glycans using 2-amino benzamide and anthranilic acid. Anal Biochem 230:229–238

Bjellqvist B, Hughes G, Pasquali C, Paquet N, Ravier F, Sanchez JC, Frutiger S, Hochstrasser D (1993) The focusing positions of polypeptides in immobilized pH gradients can be predicted from their amino acid sequences. Electrophoresis 14:1023–1031

Bjellqvist B, Basse B, Olsen E, Celis JE (1994) Reference points for comparisons of two-dimensional maps of proteins from different human cell types defined in a pH scale where isoelectric points correlate with polypeptide compositions. Electrophoresis 15:529–539

Carr SA, Hemling ME, Bean MF, Roberts GD (1991) Integration of mass spectrometry in analytical biotechnology. Anal Chem 63:2802–2824

Clauser KR, Hall SC, Smith DM, Webb JW, Andrews LE, Tran HM, Epstein LB, Burlingame AL (1995) Rapid mass spectrometric peptide sequencing and mass matching for characterization of human melanoma proteins isolated by two-dimensional PAGE. Proc Natl Acad Sci USA 92:5072–5076

Cunningham BA, Hemperly JJ, Murray BA, Prediger EA, Brackenbury R, Edelman GM (1987) Neural cell adhesion molecule: structure, immunoglobulin-like domains, cell surface modulation, and alternative RNA splicing. Science 236:799–806

Devine PL, Warren JA (1990) Glycoprotein detection on Immobilon PVDF transfer membrane using the periodic acid/Schiff reagent. BioTechniques 8:492–495

Driessen HP, De Jong WW, Tesser GI, Bloemendal H (1985) The mechanism of N-terminal acetylation of proteins. CRC Crit Rev Biochem 18:281–325

Ducret A, Bruun CF, Bures EJ, Marhaug G, Aebersold R (1996) Characterization of human serum amyloid A protein isoforms separated by two-dimensional electrophoresis by liquid chromatography/electrospray ionization tandem mass spectrometry. Electrophoresis 17:866–876

Eckerskorn C, Strupat K, Schleuder D, Hochstrasser DF, Sanchez JC, Lottspeich F, Hillenkamp F (1997) Analysis of proteins by direct scanning infrared MALDI mass spectrometry after 2D-PAGE separation and electroblotting. Anal Chem in press

Friedman Y, Higgins EA (1995) A method for monitoring the glycosylation of recombinant glycoproteins from conditioned medium, using fluorophore-assisted carbohydrate electrophoresis. Anal Biochem 228:221–225

Gianazza E (1995) Isoelectric focusing as a tool for the investigation of post-translational processing and chemical modifications of proteins. J Chromatogr A 705:67–87

Goldstein M, Lee KY, Lew JY, Harada K, Wu J, Haycock JW, Hokfelt T, Deutch AY (1995) Antibodies to a segment of tyrosine hydroxylase phosphorylated at serine 40. J Neurochem 64:2281–2287

Gooley AA, Williams KL (1997) How to find, identify and quantitate the sugars on proteins. Nature 358:557–559

Gooley AA, Ou K, Russell J, Wilkins MR, Sanchez JC, Hochstrasser DF, Williams KL (1997) A role for Edman degradation in proteome studies. Electrophoresis, in press

Grant GA, Crankshaw MW (1997) Identification of PTH-amino acids by high-performance liquid chromatography. In: Smith BJ (ed) Methods in molecular biology Vol 64. Protein sequencing protocols. Humana Press Inc, Tolowa, NJ, pp 197–215

Gravel P, Walzer C, Aubry C, Balant LP, Yersin B, Hochstrasser DF, Guimon J (1996) New alterations of serum glycoproteins in alcoholic and cirrhotic patients revealed by high resolution two-dimensional gel electrophoresis. Biochem Biophys Res Commun 220:78–85

Graves DJ, Martin BL, Wang JH (1994) Co- and post-translational modification of proteins: Chemical principles and biological effects. Oxford University Press, Oxford

Hansen JE, Lund O, Engelbrecht J, Bohr H, Nielson JO, Hansen JE, Brunak S (1995) Prediction of O-glycosylation of mammalian proteins: specificity patterns of UDP-GalNAc:polypeptide N-acetylgalactosaminyltransferase. Biochem J 308:801–813

Harvey DJ, Rudd PM, Bateman RH, Bordoli RS, Howes K, Hoyes JB, Vickers RG (1994) Examination of complex oligosaccharides by matrix-assisted laser desorption ionization mass spectrometry on time-of-flight and magnetic sector instruments. Org Mass Spectrom 29:753–765

Hellman U, Wernstedt C, Gonez J, Heldin CH (1995) Improvement of an "in-gel" digestion procedure for the micropreparation of internal protein fragments for amino acid sequencing. Anal Biochem 224:451–455

Hidaka H, Fidge NH (1992) Affinity purification of the hepatic high-density lipoprotein receptor identifies two acidic glycoproteins and enables further characterization of their binding properties. Biochem J 284:161–167

Hildebrandt E, Fried VA (1989) Phosphoaminoacid analysis of protein immobilized on polyvinylidene difluoride membrane. Anal Biochem 177:407–412

Hirano H, Komatsu S, Kajiwara H, Takagi Y, Tsunasawa S (1993) Microsequence analysis of the N-terminally blocked proteins immobilized on polyvinylidene difluoride membrane by western blotting. Electrophoresis 14:839–846

House C, Wettenhall RE, Kemp BE (1987) The influence of basic residues on the substrate specificity of protein kinase C. J Biol Chem 262:772–777

Houthaeve T, Gausepohl H, Mann M, Ashman K (1995) Automation of micro-preparation and enzymatic cleavage of gel electrophoretically separated proteins. FEBS Lett 376:91–94

Hughes GJ, Frutiger S, Paquet N, Ravier F, Pasquali C, Sanchez JC, James R, Tissot JD, Bjellqvist B, Hochstrasser D (1992) Plasma protein map: an update by microsequencing. Electrophoresis 13:707–714

James DC (1996) Analysis of recombinant glycoproteins by mass spectrometry. Cytotechnology 22:17–24

James DC, Freedman RB, Hoare M, Ogpnah OW, Rooney, BC, Larionov OA, Dobrolvolsky VN, Lagutin OV, Jenkins N (1995) N-glycosylation of recombinant human interferon-γ produced in animal expression systems. Bio/Technology 13:592–596

Johnson RD, Bhatnagar RS, Knoll LJ, Gordon JI (1994) Genetic and biochemical studies of protein N-myristoylation. Ann Rev Biochem 63:869–914

Kawashima H, Murata T, Yamamoto K, Tateishi A, Irimura T, Osawa T (1992) A simple method for the release of asparagine-linked oligosaccharides from a glycoprotein purified by SDS-polyacrylamide gel electrophoresis. J Biochem 111:620–622

Kellner R, Lottspeich F, Meyer HE (eds) (1994) Microcharacterization of proteins. VCH Weinheim

Küster B, Naven TJP, Harvey DJ (1996) Rapid approach for sequencing neutral oligosaccharides by exoglycosidase digestion and matrix-assisted laser desorption/ionization time-of-flight mass spectrometry. J Mass Spec 31:1131–1140

Lee FJ, Lin LW, Smith JA (1989) N alpha-acetyltransferase deficiency alters protein synthesis in *Saccharomyces cerevisiae*. FEBS Lett 256:139–142

Liao PC, Leykam J, Andrews PC, Gage DA, Allison J (1994) An approach to locate phosphorylation sites in a phosphoprotein: mass mapping by combining specific enzymatic degradation with matrix-assisted laser desorption/ionisation mass spectrometry. Anal Biochem 219:9–20

Lui M, Tempst P, Erdjument-Bromage H (1996) Methodical analysis of protein-nitrocellulose interactions to design a refined digestion protocol. Anal Biochem 241:156–166

Lo-Guidice JM, Wieruszeski JM, Lemoine J, Verbert A, Roussel P, Lamblin G (1994) Sialylation and sulfation of the carbohydrate chains in respiratory mucins from a patient with cystic fibrosis. J Biol Chem 269:18794–18813

Lopez-Rodas G, Georgieva EI, Sendra R, Loidl P (1991) Histone acetylation in *Zea mays* I. J Biol Chem 266:18745–18750

Matsudaira P (1987) Sequence of picomole quantities of proteins electroblotted onto polyvinylidene difluoride membranes. J Biol Chem 256:7990–7997

Medzihradszky KF, Gillececastro BL, Townsend RR, Burlingame AL, Hardy MR (1996) Structural elucidation of O-linked glycopeptides by high-energy collision-induced dissociation. 7:319–328

Meyer HE (1994) Analysing post-translational modifications. In: Kellner R, Lottspeich F, Meyer HE (eds) Microcharacterization of proteins. VCH Weinheim, pp 131–146

Mörtz E, Sareneva T, Haebel S, Julkunen I, Roepstorff P (1996) Mass spectrometric characterization of glycosylated interferon-gamma variants separated by gel electrophoresis. Electrophoresis 17:925–931

Mullen JR, Kayne PS, Moerschell RP, Tsunasawa S, Gribskov M, Colavito-Shepanski M, Grunstein-Sherman F, Sternglanz R (1989) Identification and characterisation of genes and mutants in N-terminal acetyltransferase from yeast. EMBO J 8:2067–2075

Neuberger A (1988) Early work on the glycoprotein of egg albumin. Trends Biochem Sci 13:398–399

Oh-eda M, Tominaga E, Nabuchi Y, Matsuura T, Ochi N, Tamura M, Hase S (1996) Preparation of pyridylaminated O-linked sugar chains from glycoproteins blotted on to a polyvinylidene difluoride membrane and application to human granulocyte colony-stimulating factor. Anal Biochem 236:369–371

Packer NH, Wilkins MR, Golaz O, Lawson MA, Gooley AA, Hochstrasser DF, Redmond JW, Williams KL (1996) Characterization of human plasma glycoproteins separated by two-dimensional gel electrophoresis. Bio/Technology 14:66–70

Packer NH, Ball M, Devine PL (1997) Glycoprotein detection. In: Link AJ (ed) 2-D protein gel electrophoresis protocols. Humana Press Totowa, NJ, in press

Pappin DJC, Rahman D, Hansen HF, Jeffery W, Sutton CW (1995) Peptide-mass fingerprinting as a tool for the rapid identification and mapping of cellular proteins. In: Atassi MZ, Appella E (eds) Methods in protein structure analysis. Plenum Press, New York, pp 161–174

Parekh RB, Tse AG, Dwek RA, Williams AF, Rademacher TW (1987) Tissue-specific N-glycosylation, site-specific oligosaccharide patterns and lentil lectin recognition of rat Thy-1. EMBO J 6:1233–1244

Perlman D, Halvorson HO (1983) A putative signal peptidase recognition site and sequence in eukaryotic and prokaryotic signal peptides. J Mol Biol 167:391–409

Posch A, Weiss W, Wheeler C, Dunn MJ, Görg A (1995) Sequence analysis of wheat grain allergens separated by two-dimensional electrophoresis with immobilized pH gradients. Electrophoresis 16:1115–1119

Roberts GD, Johnson WP, Burman S, Anumula KR, Carr SA (1995) An integrated strategy for structural characterization of the protein and carbohydrate components of monoclonal antibodies: application to anti-respiratory syncytial virus Mab. Anal Chem 67:3613–3625

Rucker RB, McGee C (1993) Chemical modifications of proteins in vivo: selected examples important to cellular recognition. J Nutr 123:977–990

Sutton CW, Wheeler CH, U S, Corbett JM, Cottrell JS, Dunn MJ (1997) The analysis of myocardial proteins by infrared and ultraviolet laser desorption mass spectrometry. Electrophoresis 18:424–431

Terwilliger TC, Wang JY, Koshland DE (1986) Kinetics of receptor modification. The multiply methylated aspartate receptors involved in bacterial chemotaxis. J Biol Chem 261:10814–10820

Thornton DJ, Carlstedt I, Sheehan JK (1996) Identification of glycoproteins on nitrocellulose membranes and gels. Mol Biotechnol 5:171–176

Udenfriend S, Kodukula K (1995) How glycosyl-phosphatidylinositol-anchored membrane proteins are made. Ann Rev Biochem 64:563–591

Wang JY (1988) Antibodies for phosphotyrosine: analytical and preparative tool for tyrosyl-phosphorylated proteins. Anal Biochem 172:1–7

Watts JD, Affolter M, Krebs DL, Wange RL, Samelson LE, Aebersold R (1995) Electrospray ionisation mass spectrometric investigation of signal transduction pathways. In Snyder AP (ed) Biochemical and biotechnological applications of electrospray ionization mass spectrometry. American Chemistry Society, Washington, pp 381–407

Weitzhandler M, Kadlecek D, Avdalovic N, Forte JG, Chow D, Townsend RR (1993) Monosaccharide and oligosaccharide analysis of proteins transferred to polyvinylidene fluoride membranes after sodium dodecyl sulfate-polyacrylamide gel electrophoresis. J Biol Chem 268:5121–5130

Wilkins MR, Sanchez JC, Williams KL, Hochstrasser DF (1996) Current challenges and future applications for protein maps and post-translational vector maps in proteome projects. Electrophoresis 17:830–838

Wirth PJ, Hoang TN, Benjamin T (1995) Micropreparative immobilized pH gradient two-dimensional electrophoresis in combination with protein microsequencing for the analysis of human liver proteins. Electrophoresis 16:1946–1960

Yamada R, Bradshaw RA (1991) Rat liver polysome N alpha-acetyltransferase: substrate specificity. Biochemistry 30:1017–1021

Yan JX, Packer NH, Tonella L, Ou K, Wilkins MR, Sanchez JC, Gooley AA, Hochstrasser DF, Williams KL (1997) High sample throughput phosphoamino acid analysis of proteins separated by one-and two-dimensional gel electrophoresis. J Chromatogr A 764:201–210

Zhang FL, Casey PJ (1996) Protein prenylation: Molecular mechanisms and functional consequences. Ann Rev Biochem 65:241–269

Zhang W, Czernik AJ, Yungwirth T, Aebersold R, Chait BT (1994) Matrix-assisted laser desorption mass spectrometric peptide mapping of proteins separated by two-dimensional gel electrophoresis: determination of phosphorylation in synapsin I. Protein Sci 3:677–686

Zhu Q, Bhavanandan VP (1995) Analysis of serine/threonine-linked oligosaccharides derived by alkaline-borohydride treatment of mucin glycoproteins electroblotted onto membranes: comparison of the saccharide profiles of the 390 kDa and 350 kDa forms of epinectin. Glycoconjugate J 12:639–644

5 Proteome Databases

Amos Bairoch

5.1 Introduction

The awareness that protein and DNA sequence data are essential to the understanding of biological systems is now well established in the life science community. This community is progressively becoming conscious that this is also true of additional information about protein expression, post-translational modifications, tertiary structure and, of course, function. All of this knowledge needs to be encapsulated in various databases.

The goal of this chapter is to describe the data resources that are available to researchers working in the field of proteome studies. We will not attempt here to survey all the different databases that are relevant to this field. Such an exercise would be tedious due to the large number of relevant databases and would only be valid for a very short period of time due to the extreme speed with which new databases are appearing and/or disappearing. It is also for this reason that you will find a table at the end of this chapter (Table 5.1) listing the World-Wide Web (WWW) addresses of the databases described in the following sections. The most important component of this table is the Internet address that allows you to download an up-to-date version of the table! We will successively describe the type of information found in the following types of databases: protein sequence, nucleotide sequence, pattern/profile, 2-D PAGE, 3-D structure, post-translational modification, genomic and metabolic. The last section of this chapter will try to predict future trends in the evolution of protein information resources.

5.2 Protein sequence databases

The most comprehensive source of information on proteins is found in protein sequence databases. Two types of such databases can be discerned: universal databases that strive to store information on all types of proteins from all biological species; and specialised databases that cater for either a specific group or families of proteins or for a specific organism. Universal protein sequence databases can themselves be separated into two categories: databases that are a simple repository

```
ID   ATPB_ECOLI     STANDARD;       PRT;    459 AA.
AC   P00824;
DT   21-JUL-1986 (REL. 01, CREATED)
DT   01-FEB-1995 (REL. 31, LAST SEQUENCE UPDATE)
DT   01-FEB-1997 (REL. 35, LAST ANNOTATION UPDATE)
DE   ATP SYNTHASE BETA CHAIN (EC 3.6.1.34).
GN   ATPD OR UNCD OR PAPB.
OS   ESCHERICHIA COLI.
OC   PROKARYOTA; GRACILICUTES; SCOTOBACTERIA; FACULTATIVELY ANAEROBIC RODS;
OC   ENTEROBACTERIACEAE.
RN   [1]
RP   SEQUENCE FROM N.A.
RX   MEDLINE; 85121806.
RA   WALKER J.E., GAY N.J., SARASTE M., EBERLE A.N.;
RL   BIOCHEM. J. 224:799-815(1984).
RN   [2]
RP   SEQUENCE FROM N.A.
RX   MEDLINE; 82059507.
RA   SARASTE M., GAY N.J., EBERLE A., RUNSWICK M.J., WALKER J.E.;
RL   NUCLEIC ACIDS RES. 9:5287-5296(1981).
..
...
..
RN   [8]
RP   SEQUENCE OF 1-11.
RC   STRAIN=K12 / W3110;
RA   FRUTIGER S., HUGHES G.J., PASQUALI C., HOCHSTRASSER D.F.;
RL   SUBMITTED (FEB-1996) TO THE SWISS-PROT DATA BANK.
RN   [9]
RP   MUTAGENESIS.
RX   MEDLINE; 91358411.
RA   IWAMOTO A., OMOTE H., HANADA H., TOMIOKA N., ITAI A., MAEDA M.,
RA   FUTAI M.;
RL   J. BIOL. CHEM. 266:16350-16355(1991).
CC   -!- FUNCTION: PRODUCES ATP FROM ADP IN THE PRESENCE OF A PROTON
CC       GRADIENT ACROSS THE MEMBRANE. THE BETA CHAIN IS THE CATALYTIC
CC       SUBUNIT.
CC   -!- SUBUNIT: F-TYPE ATPASES HAVE 2 COMPONENTS, CF(1) - THE CATALYTIC
CC       CORE - AND CF(0) - THE MEMBRANE PROTON CHANNEL. CF(1) HAS FIVE
CC       SUBUNITS: ALPHA(3), BETA(3), GAMMA(1), DELTA(1), EPSILON(1). CF(0)
CC       HAS THREE MAIN SUBUNITS: A, B AND C.
CC   -!- SIMILARITY: BELONGS TO THE ATPASE ALPHA/BETA CHAINS FAMILY.
DR   EMBL; X01631; G899257; -.
DR   EMBL; V00267; G41038; -.
DR   EMBL; M25464; G146325; -.
DR   EMBL; J01594; G148139; -.
DR   EMBL; V00311; G42277; -.
DR   EMBL; V00312; G42285; -.
DR   EMBL; L10328; G290581; -.
DR   EMBL; AE000450; G1790170; -.
DR   PIR; A01023; PWECB.
DR   PIR; S02748; S02748.
DR   SWISS-2DPAGE; P00824; COLI.
DR   ECO2DBASE; B046.7; 6TH EDITION.
DR   ECOGENE; EG10101; ATPD.
DR   PROSITE; PS00152; ATPASE_ALPHA_BETA.
KW   HYDROLASE; ATP SYNTHESIS; CF(1); ATP-BINDING; HYDROGEN ION TRANSPORT.
FT   INIT_MET      0      0
FT   NP_BIND     149    156       ATP (BY SIMILARITY).
FT   MUTAGEN     156    156       T->A,C: IMPAIRS ATPASE ACTIVITY.
SQ   SEQUENCE   459 AA;  50194 MW;  2B50A863 CRC32;
     ATGKIVQVIG AVVDVEFPQD AVPRVYDALE VQNGNERLVL EVQQQLGGGI VRTIAMGSSD
     GLRRGLDVKD LEHPIEVPVG KATLGRIMNV LGEPVDMKGE IGEEERWAIH RAAPSYEELS
     NSQELLETGI KVIDLMCPFA KGGKVGLFGG AGVGKTVNMM ELIRNIAIEH SGYSVFAGVG
     ERTREGNDFY HEMTDSNVID KVSLVYGQMN EPPGNRLRVA LTGLTMAEKF RDEGRDVLLF
     VDNIYRYTLA GTEVSALLGR MPSAVGYQPT LAEEMGVLQE RITSTKTGSI TSVQAVYVPA
     DDLTDPSPAT TFAHLDATVV LSRQIASLGI YPAVDPLDST SRQLDPLVVG QEHYDTARGV
     QSILQRYQEL KDIIAILGMD ELSEEDKLVV ARARKIQRFL SQPFFVAEVF TGSPGKYVSL
     KDTIRGFKGI MEGEYDHLPE QAFYMVGSIE EAVEKAKKL
//
```

Fig. 5.1. Excerpt of a sample SWISS-PROT entry

of sequence data, mostly translated directly from DNA sequences; and annotated databases where information other than the sequence is extracted by biologists (known as annotators) from the original literature, review articles as well as other electronic archives. Here we will describe in detail SWISS-PROT which is an annotated universal sequence database, TrEMBL which is a supplement of SWISS-PROT that can be classified as a repository of sequence data, and we will also discuss in detail the issues of completeness and non-redundancy. Finally we will deal with a few examples of specialised protein sequence databases.

5.2.1 SWISS-PROT

Introduction. SWISS-PROT (Bairoch and Apweiler 1997; Apweiler et al. 1997) is a curated protein sequence database which strives to provide a high level of annotations (such as the description of the function of a protein, its domain structure, post-translational modifications and variants), a minimal level of redundancy, a high level of integration with other biomolecular databases as well as extensive external documentation.

It was created in 1986 and since 1988 it has been a collaborative endeavour of the Department of Medical Biochemistry of the University of Geneva and the EMBL Data Library (now at the EMBL Outstation — The European Bioinformatics Institute (EBI) (Robinson 1994)). SWISS-PROT contains data that originates from a wide variety of biological organisms; currently there are a total of about 65,000 annotated entries from more than 5,000 different species. But the bulk of entries comes from about 30 species that are the target of many biological studies. The ten top species in SWISS-PROT ranked by number of entries are: human, the budding yeast *Saccharomyces cerevisiae*, *Escherichia coli*, mouse, rat, *Bacillus subtilis*, *Haemophilus influenzae*, *Caenorhabditis elegans*, the fission yeast *Schizosaccharomyces pombe* and the fruit fly *Drosophila melanogaster*. A sample SWISS-PROT entry is shown in Fig. 5.1.

The SWISS-PROT format. As you can see from Fig. 5.1, a SWISS-PROT entry is made up of different types of lines. Each line type begins with a two-character line code, which indicates the type of data contained in the line. Each line type has its own format used to record the various data which make up an entry. There are 21 different line types in SWISS-PROT. Some entries do not contain all of the line types, and some line types may occur many times in a single entry. Each entry begins with an identification line (ID) and ends with a terminator line (//).

A detailed look at a SWISS-PROT entry. The best way to understand the type of information stored in SWISS-PROT is to look in detail at the example shown in Fig. 5.1 and to explain the type of information which is found on the different line types. We should first look at the beginning of the entry:

```
ID   ATPB_ECOLI      STANDARD;       PRT;    459 AA.
AC   P00824;
DT   21-JUL-1986 (REL. 01, CREATED)
DT   01-FEB-1995 (REL. 31, LAST SEQUENCE UPDATE)
DT   01-FEB-1997 (REL. 35, LAST ANNOTATION UPDATE)
DE   ATP SYNTHASE BETA CHAIN (EC 3.6.1.34).
GN   ATPD OR UNCD OR PAPB.
OS   ESCHERICHIA COLI.
OC   PROKARYOTA; GRACILICUTES; SCOTOBACTERIA; FACULTATIVELY ANAEROBIC RODS;
OC   ENTEROBACTERIACEAE.
```

The first line of each SWISS-PROT entry is the ID (IDentification) line. It contains the entry name (here 'ATPB_ECOLI') which is a useful handle to retrieve this particular entry in the database. But if you need to cite a particular entry in a publication, it is much better to use the code which is shown on the second line, the AC (ACcession) line. This code (here 'P00824') is known as the accession number. It is a stable identifier while an entry name can change over time. It sometimes happens that the AC line contains more than one accession number, in this case you should always cite the first one which is known as the "primary accession number". The AC line is followed by three DT (DaTe) lines which tell you when the sequence was first entered in SWISS-PROT, when the sequence itself was last modified and when the most recent information was added to the entry. We then encounter the DE (DEscription) line(s) which give you the name(s) of the protein. An effort is made to list all the names under which a particular protein is or has been known. This can sometimes be quite extensive as shown in an example from the entry for human annexin V:

```
DE   ANNEXIN V (LIPOCORTIN V) (ENDONEXIN II) (CALPHOBINDIN I) (CBP-I)
DE   (PLACENTAL ANTICOAGULANT PROTEIN I) (PAP-I) (PP4) (THROMBOPLASTIN
DE   INHIBITOR) (VASCULAR ANTICOAGULANT-ALPHA) (VAC-ALPHA) (ANCHORIN CII).
```

Returning to ATPB_ECOLI, the next line, GN (Gene Name) lists the name(s) of the protein's gene. This line can be absent if no gene name has been assigned or it can contain more than one name if multiple designations have been given for a gene by different groups or at different periods in time.

We then arrive at the lines that describe the biological organism from which the protein is derived. These are the OS (Organism Species) and OC (Organism Classification) lines. The OS line gives the organism's name in Latin — first its genus then the species it belongs to — and, if available, its name in English. The OC lines contain the organism's classification within the various biological kingdoms. Such a classification is known as a taxonomic tree or simply as "the tree of life". The OC lines are particularly useful if you want to extract a subset of the database based on taxonomic criteria or if you want to restrict a protein identification tool to only deal with a specific subset (see Sect. 3.3.1).

A third line, which is not present in the above example is the OG (OrGanelle) line. If appropriate, this is used to indicate in what organelle or extra-chromosomal element the sequenced gene is encoded, such as a chloroplast, a cyanelle, a mitochondrion or a plasmid. Example:

```
OG   MITOCHONDRION.
```

Let us now look at the next part of the sample SWISS-PROT entry:

```
RN   [1]
RP   SEQUENCE FROM N.A.
RX   MEDLINE; 85121806.
RA   WALKER J.E., GAY N.J., SARASTE M., EBERLE A.N.;
RL   BIOCHEM. J. 224:799-815(1984).
..
...
..
RN   [8]
RP   SEQUENCE OF 1-11.
RC   STRAIN=K12 / W3110;
RA   FRUTIGER S., HUGHES G.J., PASQUALI C., HOCHSTRASSER D.F.;
RL   SUBMITTED (FEB-1996) TO THE SWISS-PROT DATA BANK.
RN   [9]
RP   MUTAGENESIS.
RX   MEDLINE; 91358411.
RA   IWAMOTO A., OMOTE H., HANADA H., TOMIOKA N., ITAI A., MAEDA M.,
RA   FUTAI M.;
RL   J. BIOL. CHEM. 266:16350-16355(1991).
```

This section contains various references. Each reference is organised as a block of lines starting with 'R': RN, RP, RX, RC, RA and RL. The RN line — 'N' for number — is simply the number of the reference in an entry. The RP line is used to very briefly mention the type of work which has been carried out on a protein. The RC line — 'C' for comment — is used to indicate information such as the tissue from which the protein was extracted ('TISSUE=') or as in the example above, from what strain ('STRAIN='). The RX line — 'X' for cross-reference — is used to indicate the identifier assigned to a specific reference in a bibliographic database such as Medline. The use of the RA line — 'A' for author — is straightforward as is the RL line — 'L' for location — which contains the conventional citation information for the reference.

A number of things should be mentioned here:

1. As can be seen from reference 8 in the example above, SWISS-PROT contains not only references to published journal articles, books and theses but also references to information directly submitted to the database. In fact there are increasing numbers of sequences that are never "published" except in databases; and the same trend is true for protein characterisation data coming from proteome projects. It is important to be aware of this trend and to realise that biomolecular databases are becoming much more than a repository of information that can be found elsewhere.

2. SWISS-PROT entries include not only references about sequencing work, but also references for a large variety of other type of studies such as 3-D structure determination, mutagenesis (see reference 9 in the sample entry), determination of variants and detection of post-translational modifications.

3. SWISS-PROT does not store the title of a reference; however this will change in the near future with titles being present in a new RT (Reference Title) line.

Continuing in the sample entry we arrive at the following section:

```
CC   - ! -  FUNCTION: PRODUCES ATP FROM ADP IN THE PRESENCE OF A PROTON
CC          GRADIENT ACROSS THE MEMBRANE. THE BETA CHAIN IS THE CATALYTIC
CC          SUBUNIT.
CC   - ! -  SUBUNIT: F-TYPE ATPASES HAVE 2 COMPONENTS, CF(1) - THE CATALYTIC
CC          CORE - AND CF(0) - THE MEMBRANE PROTON CHANNEL. CF(1) HAS FIVE
CC          SUBUNITS: ALPHA(3), BETA(3), GAMMA(1), DELTA(1), EPSILON(1). CF(0)
CC          HAS THREE MAIN SUBUNITS: A, B AND C.
CC   - ! -  SIMILARITY: BELONGS TO THE ATPASE ALPHA/BETA CHAINS FAMILY.
```

The CC (Comments) lines contain a variety of textual comments which are classed into different "topics". In the above example, data has been classed into the different topics of 'FUNCTION', 'SUBUNIT' and 'SIMILARITY'. There are all together 19 different types of topics. Some of them, like 'PTM' or 'MASS SPECTROMETRY' are particularly relevant to proteome studies. Examples:

```
CC   - ! -  PTM: ACYLATED, PHOSPHORYLATED ON SERINE AND N-GLYCOSYLATED WITH
CC          TWO TYPES OF OLIGOSACCHARIDE CHAINS.
CC   - ! -  MASS SPECTROMETRY: MW=71890; MW_ERR=7; METHOD=ELECTROSPRAY.
```

Let's continue with the next section from the sample entry:

```
DR   EMBL; X01631; G899257; -.
DR   EMBL; V00267; G41038; -.
DR   EMBL; M25464; G146325; -.
DR   EMBL; J01594; G148139; -.
DR   EMBL; V00311; G42277; -.
DR   EMBL; V00312; G42285; -.
DR   EMBL; L10328; G290581; -.
DR   EMBL; AE000450; G1790170; -.
DR   PIR; A01023; PWECB.
DR   PIR; S02748; S02748.
DR   SWISS-2DPAGE; P00824; COLI.
DR   ECO2DBASE; B046.7; 6TH EDITION.
DR   ECOGENE; EG10101; ATPD.
DR   PROSITE; PS00152; ATPASE_ALPHA_BETA.
```

The DR (Database cross-Reference) line provides links (cross-references) from SWISS-PROT to other biomolecular databases. A detailed description of the DR line is available in Sect. 6.3.1. In the above example, cross-references are available to six different databases. SWISS-PROT is currently linked to a total of 28 different databases.

After the DR lines comes:

```
KW   HYDROLASE; ATP SYNTHESIS; CF(1); ATP-BINDING; HYDROGEN ION TRANSPORT.
```

The KW (KeyWord) lines list relevant keywords that can be used to retrieve a specific subset of the database.

We now arrive at:

```
FT   INIT_MET       0       0
FT   NP_BIND      149     156     ATP (BY SIMILARITY).
FT   MUTAGEN      156     156     T->A,C: IMPAIRS ATPASE ACTIVITY.
```

These are the FT (FeaTure) lines which collectively comprise what is known as the feature table of an entry. The feature table provides a precise but simple means for the annotation of sequence data, describing regions or sites of interest in the sequence. In general the feature table lists post-translational modifications, binding sites, enzyme active sites, local secondary structure or other characteristics. Sequence conflicts between references are also included in the feature table. A feature line consist of three parts: a feature key (e.g. 'NP_BIND'), a position in the sequence ('149   156') and a comment part ('ATP (BY SIMILARITY)').

Here is a second example of a the feature table from the SWISS-PROT entry for human arylsulfatase A:

```
FT   SIGNAL        1     18
FT   CHAIN        19    507    ARYLSULFATASE A.
FT   CHAIN        19    444    COMPONENT B.
FT   CHAIN       448    507    COMPONENT C.
FT   MOD_RES      69     69    2-AMINO-3-OXOPROPIONIC ACID.
FT   ACT_SITE    125    125    POTENTIAL.
FT   CARBOHYD    158    158    POTENTIAL.
FT   CARBOHYD    184    184    POTENTIAL.
FT   CARBOHYD    350    350
FT   VARIANT      82     82    P -> L (IN MLD; LATE-INFANTILE ONSET).
```

In the above example there are three lines with the feature key 'CARBOHYD' that indicate positions where a carbohydrate side chain is attached to the protein backbone. Two of these lines have the comment 'POTENTIAL' while the third one does not. This signifies that in the first two cases the presence of a carbohydrate has not been experimentally proven but only inferred through sequence analysis. One can therefore distinguish between experimentally determined data and either computational predictions or features added by comparison with another entry (which are tagged with the comment 'BY SIMILARITY').

To finish our in-depth exploration of the SWISS-PROT entry ATPB_ECOLI, we arrive at:

```
SQ   SEQUENCE   459 AA;   50194 MW;   2B50A863 CRC32;
     ATGKIVQVIG AVVDVEFPQD AVPRVYDALE VQNGNERLVL EVQQQLGGGI VRTIAMGSSD
     GLRRGLDVKD LEHPIEVPVG KATLGRIMNV LGEPVDMKGE IGEEERWAIH RAAPSYEELS
     NSQELLETGI KVIDLMCPFA KGGKVGLFGG AGVGKTVNMM ELIRNIAIEH SGYSVFAGVG
     ERTREGNDFY HEMTDSNVID KVSLVYGQMN EPPGNRLRVA LTGLTMAEKF RDEGRDVLLF
     VDNIYRYTLA GTEVSALLGR MPSAVGYQPT LAEEMGVLQE RITSTKTGSI TSVQAVYVPA
     DDLTDPSPAT TFAHLDATVV LSRQIASLGI YPAVDPLDST SRQLDPLVVG QEHYDTARGV
     QSILQRYQEL KDIIAILGMD ELSEEDKLVV ARARKIQRFL SQPFFVAEVF TGSPGKYVSL
     KDTIRGFKGI MEGEYDHLPE QAFYMVGSIE EAVEKAKKL
//
```

This section consists of a SQ (SeQuence header) line and the sequence itself (which is stored in one-letter amino acid code by lines of 60 residues).

It is important to note here that sequence data in SWISS-PROT corresponds to the precursor form of proteins before post-translational modifications and processing. Therefore the information about the size (here 459 amino acids) or molecular weight (here 50,194 Da) does not necessarily correspond to the values for a mature protein. Programs that identify proteins on the basis of molecular weight, isoelectric points or peptide fingerprints must take into account the feature table of SWISS-PROT entries which contains the necessary information to compute the extent of the mature proteins or peptides. For example, features such as 'SIGNAL' (see the feature table example above), 'TRANSIT' and 'PROPEP' indicate the region of the sequence which belongs respectively to a signal sequence, a transit peptide or a propeptide. The features 'CHAIN' or 'PEPTIDE' indicate the regions which are present in the mature form. Most protein identification tools on the ExPASy WWW server (Appel et al. 1994) consider annotation data from SWISS-PROT feature tables (see Chap. 3).

5.2.2 TrEMBL

The ongoing genome sequencing and mapping projects have dramatically increased the number of protein sequences to be incorporated into SWISS-PROT. Since it was not desirable to dilute the quality standards of SWISS-PROT by incorporating sequences without proper sequence analysis and annotation, but as it is necessary for many applications such as sequence similarity searches and proteome projects to be able to access all known sequences as quickly as possible, it was decided in 1996 to create TrEMBL. TrEMBL (TRanslation of EMBL nucleotide sequence database) is a protein sequence database supplementing SWISS-PROT. It consists of computer-annotated entries in SWISS-PROT-like format derived from the translation of coding sequences (CDS) in the EMBL nucleotide sequence database (see Sect. 5.3). TrEMBL can be considered as a preliminary section of SWISS-PROT, where sequences are kept before being manually annotated and moved to the main part of the SWISS-PROT database.

As it can be seen from the sample TrEMBL entry shown in Fig. 5.2, the format of TrEMBL is identical to that of SWISS-PROT. There are generally not many annotations in a TrEMBL entry other than those that can be derived from the source EMBL nucleotide sequence entry and those that can be added automatically by computer software. In TrEMBL the entry names are identical to the primary accession numbers.

5.2.3 Completeness and non-redundancy issues

For many applications such as sequence similarity searches or protein identification (see Chap. 3) it is desirable to be able to scan a database which is as complete as possible and, hopefully, also non-redundant. The combination of SWISS-PROT and TrEMBL offers a solution to both these requirements. As we will see later, this solution is not yet optimal but is a considerable improvement on those previously available. Because the databases which were created by these previous attempts are still used as the core of many identification and search services on the WWW, we will describe them here. For a number of years the NCBI (National Center for Biotechnology Information) has offered a database called 'nrdb' (Non-Redundant DataBase) as the default option of their BLAST (Altschul et al. 1990) similarity search server. Nrdb is built from the addition of SWISS-PROT and GenPept (an automatic translation of the GenBank™ nucleotide sequence database). The criteria used to exclude sequences from nrdb is very primitive in that it is based only on

```
ID   Q12841        PRELIMINARY;    PRT;    308 AA.
AC   Q12841;
DT   01-NOV-1996 (TREMBLREL. 01, CREATED)
DT   01-NOV-1996 (TREMBLREL. 01, LAST SEQUENCE UPDATE)
DT   01-NOV-1996 (TREMBLREL. 01, LAST ANNOTATION UPDATE)
DE   FOLLISTATIN-RELATED PROTEIN PRECURSOR.
OS   HOMO SAPIENS (HUMAN).
OC   EUKARYOTA; METAZOA; CHORDATA; VERTEBRATA; TETRAPODA; MAMMALIA;
OC   EUTHERIA; PRIMATES.
RN   [1]
RP   SEQUENCE FROM N.A.
RC   TISSUE=BRAIN;
RX   MEDLINE; 95045570.
RA   ZWIJSEN A., BLOCKX H., VAN ARNHEM W., WILLEMS J., FRANSEN L.,
RA   DEVOS K., RAYMACKERS J., DE VOORDE A., SLEGERS H.;
RL   EUR. J. BIOCHEM. 225:937-946(1994).
DR   EMBL; U06863; G536898; -.
KW   SIGNAL.
FT   SIGNAL         1      20        POTENTIAL.
FT   CHAIN         21     308        FOLLISTATIN-RELATED PROTEIN.
FT   CARBOHYD     144     144        POTENTIAL.
FT   CARBOHYD     175     175        POTENTIAL.
FT   CARBOHYD     180     180        POTENTIAL.
SQ   SEQUENCE    308 AA;   34985 MW;   E35CA0FA CRC32;
     MWKRWLALAL ALVAVAWVRA EEELRSKSKI CANVFCGAGR ECAVTEKGEP TCLCIEQCKP
     HKRPVCGSNG KTYLNHCELH RDACLTGSKI QVDYDGHCKE KKSVSPSASP VVCYQSNRDE
     LRRRIIQWLE AEIIPDGWFS KGSNYSEILD KYFKNFDNGD SRLDSSEFLK FVEQNETAIN
     ITTYPDQENN KLLRGLCVDA LIELSDENAD WKLSFQEFLK CLNPSFNPPE KKCALEDETY
     ADGAETEVDC NRCVCACGNW VCTAMTCDGK NQKGAQTQTE EEMTRYVQEL QKHQETAEKT
     KRVSTKEI
//
```

Fig. 5.2. A sample TrEMBL entry

the exact match between two protein sequences. Applying such a criterion results in a number of severe drawbacks: there are multiple copies of the same protein when those copies differ due to sequencing errors and/or polymorphisms; there are many erroneous sequences included which have been corrected or excluded from SWISS-PROT (but which get retranslated from the DNA database); and finally there are separate entries for fragments of existing complete sequences. The overall result is a database that while being much more complete than SWISS-PROT is highly redundant and contains a huge number of errors. The OWL database (Bleasby et al. 1994) is built on the same principle and suffers therefore from the same weaknesses. The combination of SWISS-PROT and TrEMBL is a step forward toward a complete but non-redundant protein sequence database. Yet, while it is considerably less redundant and less erroneous than nrdb and OWL, it can not be considered as truly non-redundant. We estimate that, considering the 65,000 SWISS-PROT entries and the 110,000 TrEMBL entries available in June 1997, the total number of unique proteins is probably closer to 120,000 than to the 175,000 obtained by addition of the two components. Further reduction of redundancy will either require the development of expert systems and/or a greater level of human intervention.

5.2.4 Specialised protein sequence databases

There are many specialised protein sequence databases. Some of them are quite
small and only contain a handful of entries, some are wider in scope and larger in
size. As this category of databases is quite changeable, any list we provide here
would soon be obsolete. However, in Table 5.1 we provide an address to part of a
WWW document that lists information sources for molecular biologists, which we
will keep constantly up-to-date.

We will briefly describe here three representative examples of specialised protein
sequence databases.

YPD. YPD (Payne and Garrels 1997) is a database for the proteins of *S. cerevisiae*.
It describes each protein in that organism by a number of data categories such as
the calculated molecular weight, isoelectric point, codon bias, subcellular localisa-

```
|===============================================================================|
| ASPARTATE AMINOTRANSFERASE, MITOCHONDRIAL                                     |
|-------------------------------------------------------------------------------|
| YPD NAME:       AAT1                                                           |
| SP/LISTA NAME:  AAT1                                                           |
| SGD NAME:       AAT1                                                           |
| SYNONYM LIST:   AAT1/YKL461/YKL106W                                            |
|-------------------------------------------------------------------------------|
| GENBANK   #:   Z28106x1            CHROMOSOME:          XI                     |
| PIR       #:   S37933              INTRONS:             no                     |
| SWISSPROT #:   Q01802              KNOCKOUT:                                    |
| YEPD      #:                                                                   |
|-------------------------------------------------------------------------------|
| PI:            8.970               SUBCELLULAR LOC:     mit                    |
| MOL WGT:       49360               MOLEC ENVIRONMENT:   sol                    |
| CODON BIAS:    0.087               FUNCTION CATEGORY:                          |
| CAI:           0.140                                                           |
| LEN:           431                                                            |
|-------------------------------------------------------------------------------|
| N-term modif:                      CDC28 sites, potential:  1                  |
| C-term modif:                      CKII sites, potential:   3                  |
| Phosphorylation:  unknown          PKA sites, potential:    8                  |
| Glycosylation:    unknown          N-glc sites, potential:  3                  |
| Precursor len:    20               TM domains, potential:   1                  |
|-------------------------------------------------------------------------------|
| N-term seq (precursor):   mlrtrlt                                              |
| N-term seq (mature):      srvprap                                             |
| C-term seq (mature):      skmdkla                                             |
|-------------------------------------------------------------------------------|
| Refs:  1819 20616                                                             |
|===============================================================================|
```

1819. Morin PJ; Subramanian GS; Gilmore TD. AAT1, a gene
       encoding a mitochondrial aspartate aminotransferase in
       Saccharomyces cerevisiae. Biochim Biophys Acta 1171, 211-214 (1992).
20616. Cheret G; Pallier C; Valens M; Diagnan-Fornier B;
       Fukuhara H; Bolotin-Fukuhara M; Sor F. The DNA sequence
       analysis of the HAP4-LAP4 region on chromosome XI of
       Saccharomyces cerevisiae suggests the presence of a second
       aspartate aminotransferase gene in yeast. Yeast 9, 1259-1265 (1993).

Fig. 5.3. A sample YPD entry

tion or post-translational modifications. The database is available in several formats, including a spreadsheet format as well as an easy to read report-type format as shown in Fig. 5.3. The information found in a YPD entry is similar to that found in the corresponding SWISS-PROT entry with a number of differences: YPD provides data about the gene coding for a protein (e.g. chromosome, introns, CAI (Codon Adaptation Index), codon bias) which is not found in SWISS-PROT. It also reports information on post-translational modifications and transmembrane domains in a numerical form rather than in the form of a feature table. This makes it easier to select entries according to precise criteria — for example if you want all proteins with five transmembrane regions — but with the disadvantage of losing the position and precise description of these features. Altogether YPD is a very useful data collection for all yeast researchers and especially those working on its proteome.

GCRDb. GCRDb (Kolakowski 1994) is a database of sequences and other data relevant to the study of the biology of G-protein coupled receptors (GCRs), a very large protein family of critical components of many different autocrine, paracrine and endocrine signalling systems in animals. As can be seen in Fig. 5.4, the information available in a GCRDb entry is quite succinct and is not much more extensive than what you would find in the EMBL or GenBank™ nucleotide sequence entry from which it is derived. But what makes this database useful are not the entries themselves, but the analyses (e.g. multiple alignments, classification into subfamilies) which have been made on the data and which are available from the GCRDb database. The take-home message from this example is that while some of the specialised databases often seem to provide sequence data which can be thought of being less annotated than that found in SWISS-PROT, they often offer an analytical view of the data which a universal sequence database is unable to provide.

AMSdb. AMSdb is database of antimicrobial peptides and proteins. It is oriented mainly towards animal and plant sequences but also contains some entries for proteins of bacterial origin. As can be seen from Fig. 5.5, the entries in AMSdb are derived from the corresponding entries in SWISS-PROT but are complemented by information (the line-types CY, TY and AY) specific to the subject of that database. AMSdb is an example of a new category of specialised databases that make use of the information stored in SWISS-PROT but that add a "cassette" of data specific to their field of research.

5.3 Nucleotide sequence databases

It may seem paradoxical to discuss nucleotide sequence databases in a chapter dealing with protein data resources, especially after having seen in the preceding

```
###GCR_0170    RATB2BRA    M59967    396    BRADYKININ    PcDNA
DATE_ADDED     930711
DATE_MOD       930711
FAMILY         A
GROUP          4
IUPHAR         B2-Bradykinin receptor
ABBREV         B2BKR
SPECIES Rattus norvegicus [Strain Sprague-Dawley] (Rat)
REF%K Rattus norvegicus B2 bradykinin receptor mRNA, complete cds.
REF%K GB M59967
REF%K G-protein coupled receptor; bradykinin B2 receptor;
REF%K Rattus norvegicus (strain Sprague-Dawley) cDNA to mRNA.
REF%A A. E. McEachern
REF%A E. R. Shelton
REF%A S. Bhakta
REF%A R. Obernolte
REF%A C. Bach
REF%A P. Zuppan
REF%A J. Fujisaki
REF%A R. W. Aldrich
REF%A K. Jarnagin
REF%T Expression cloning of a rat B2 bradykinin receptor
REF%J Proc. Natl. Acad. Sci. U.S.A.
REF%V 88
REF%P 7724-7728
REF%D 1991
PEP>RATB2BRA      M59967, 396 bases, 38AFAF53 checksum.
PEPMDTRSSLCPKTQAVVAVFWGPGCHLSTCIEMFNITTQALGSAHNGTFSEV
PEPNCPDTEWWSWLNAIQAPFLWVLFLLAALENIFVLSVFCLHKTNCTVAEIY
PEPLGNLAAADLILACGLPFWAITIANNFDWLFGEVLCRVVNTMIYMNLYSSI
PEPCFLMLVSIDRYLALVKTMSMGRMRGVRWAKLYSLVIWSCTLLLSSPMLVF
PEPRTMKDYREEGHNVTACVIVYPSRSWEVFTNMLLNLVGFLLPLSIITFCTV
PEPRIMQVLRNNEMKKFKEVQTEKKATVLVLAVLGLFVLCWFPFQISTFLDTL
PEPLRLGVLSGCWNERAVDIVTQISSYVAYSNSCLNPLVYVIVGKRFRKKSRE
PEPVYQAICRKGGCMGESVQMENSMGTLRTSISVDRQIHKLQDWAGNKQ
ORF290..1480/note="longestORF;NCBIgi:456683."
ORF380..1480/note="putative;NCBIgi:456684."
DNAgaattccgtt gctgtcggggg actccctaca acacagaacc ggctctctag agaaaaggac
DNAgatcctcact cgtctttgtc ctgagtacaa atgcactgtt cttggaagcg acccgtgctc
DNActgtccgtgc atgagcccat gcccaccaca gcctctctgg ggggtccaga aacactaagc
DNAaactgtccaa ggccataccg tgaccaggac ttgaccccag gaggactggt ccaaagcctc
DNAgctgtgcctc ctcccgtgca ccctctgctc acctcctgtg ctctgctgca tggacacaag
..
...
..
DNAgggacgcggc ctgggtccct cagggtggct gaagggtctg aaactgggtc actgcagggt
DNActgcaactag gtctgcagga gggaagcccc cttgtgcctt tcgtcgacac cctgatgaga
DNAaaccatgtcc attgaaggac ggagaaagct ctgtaccgtg cagaggaagt ggccgacaag
DNAggaggagcac tctgtgtaag aacacacctc ctctgtgcgc tggaataaac agatgaattt
DNAtgaaaaaaaa aaaaaaaaaa aaagcggccg c
EXN unknown
CHR unknown
TYP
```

Fig. 5.4. Excerpt of a sample GCRDb entry

section that the combination of SWISS-PROT and TrEMBL should provide a complete set of DNA-derived protein sequences. However what we have not yet discussed is that DNA databases often contain hidden treasures useful to many types of proteome studies. Before we embark on a description of these treasures it is nec-

```
ID   SR1D_SARPE    STANDARD;    PRT;    40 AA
AC   P18312;
DE   SARCOTOXIN ID.
OS   SARCOPHAGA PEREGRINA (FLESH FLY).
RP   SEQUENCE
RC   TISSUE=HEMOLYMPH
RA   MATSUYAMA K., NATORI S.;
RL   J. BIOL. CHEM. 263:17112-17116(1988).
CY   PEPTIDE WITH NO CYS
TY   ALPHA-HELICAL.
FY   CECROPINS (DROSOPHILA CECROPIN-LIKE)
AY   ANTIBACTERIAL: G+
AY   ANTIBACTERIAL: G- (E.COLI).
CC   -!- FUNCTION: SARCOTOXINS, WHICH ARE POTENT BACTERICIDAL PROTEINS,
CC            ARE PRODUCED IN RESPONSE TO INJURY. THEY ARE CYTOTOXIC
CC            TO BOTH GRAM-POSITIVE AND GRAM-NEGATIVE BACTERIA.
CC   -!- SIMILARITY: A MEMBER OF THE CECROPIN FAMILY
KW   INSECT IMMUNITY; ANTIBIOTIC; HEMOLYMPH; MULTIGENE FAMILY.
SQ   SEQUENCE  40 AA;  4349 MW;  6653 CN;
SR1D_SARPE
gwirdfgkriervgqhtrdatiqtiavaqqaanvaatlkg*
```

Fig. 5.5. A sample AMSdb entry

essary to clarify a few points about DNA databases and to introduce a number of definitions.

1. There is an international collaboration for the production, maintenance and distribution of a single resource for nucleotide sequences. For historical reasons this database is distributed under three different names and in slightly different formats by the organisations that participate in this collaboration. These organisations are the EBI (already mentioned in the context of SWISS-PROT and TrEMBL) which distributes the EMBL Nucleotide Sequence Database (Stoesser et al. 1997); the NCBI in the USA which distributes GenBank™ (Benson et al. 1997) and the National Institute of Genetics in Japan which distributes DDBJ (DNA DataBase of Japan) (Tateno and Gojobori 1997).

2. As stated above, the format of these nucleotide sequence databases is not uniform, but they share the same organisational principles as described for SWISS-PROT: a header section containing information on the name of the sequence, the species of origin followed by the references, a feature table and the sequence data. We show in Fig. 5.6 an example of a sequence entry from the EMBL database.

3. The principle of these databases is to provide an archive or repository of nucleotide sequences. By this we mean that they do not attempt to reduce the redundancy of the database by merging sequences submitted by different groups that should correspond to the same stretch of genome or mRNA. It also means that each entry is under the responsibility of the group that contributed that sequence and no attempt is made by the database staff to complement the original information with new data or the results of additional analysis. This is in stark contrast with the curation efforts made to produce SWISS-PROT.

```
ID   HSLDHAR     standard; RNA; HUM; 1661 BP.
AC   X02152;
NI   g34312
XX
DT   28-JAN-1986 (Rel. 08, Created)
DT   12-SEP-1993 (Rel. 36, Last updated, Version 2)
XX
DE   Human mRNA for lactate dehydrogenase-A (LDH-A, EC 1.1.1.27)
XX
KW   lactate dehydrogenase.
XX
OS   Homo sapiens (human)
OC   Eukaryotae; mitochondrial eukaryotes; Metazoa; Chordata;
OC   Vertebrata; Eutheria; Primates; Catarrhini; Hominidae; Homo.
XX
RN   [1]
RP   1-1661
RX   MEDLINE; 85127030.
RA   Tsujibo H., Tiano H.F., Li S.S.;
RT   "Nucleotide sequences of the cDNA and an intronless pseudogene for
RT   human lactate dehydrogenase-A isozyme";
RL   Eur. J. Biochem. 147:9-15(1985).
XX
DR   GDB; 120141; LDHA.
DR   SWISS-PROT; P00338; LDHM_HUMAN.
XX
FH   Key             Location/Qualifiers
FT   source          1..1661
FT                   /organism="Homo sapiens"
FT   CDS             98..1096
FT                   /note="lactate dehydrogenase-A"
FT                   /db_xref="PID:g34313"
FT                   /db_xref="SWISS-PROT:P00338"
FT                   /translation="MATLKDQLIYNLLKEEQTPQNKITVVGVGAVGMACAISILMKDL/
FT                   DELALVDVIEDKLKGEMMDLQHGSLFLRTPKIVSGKDYNVTANSKLVIITAGARQQEGE
FT                   SRLNLVQRNVNIFKFIIPNVVKYSPNCKLLIVSNPVDILTYVAWKISGFPKNRVIGSG(
FT                   NLDSARFRYLMGERLGVHPLSCHGWVLGEHGDSSVPVWSGMNVAGVSLKTLHPDLGTDH
FT                   DKEQWKEVHKQVVESAYEVIKLKGYTSWAIGLSVADLAESIMKNLRRVHPVSTMIKGL\
FT                   GIKDDVFLSVPCILGQNGISDLVKVTLTSEEEARLKKSADTLWGIQKELQF"
FT   misc_feature    1644..1649
FT                   /note="put. polyA signal"
FT   polyA_site      1661..1661
FT                   /note="polyA site"
SQ   Sequence 1661 BP; 458 A; 340 C; 388 G; 475 T; 0 other;
     tgctgcagcc gctgccgccg attccggatc tcattgccac gcgcccccga cgaccgcccg        6(
     acgtgcattc ccgattcctt ttggttccaa gtccaatatg gcaactctaa aggatcagct       12(
     gatttataat cttctaaagg aagaacagac ccccagaat aagattacag ttgttggggt        18(
     tggtgctgtt ggcatggcct gtgccatcag tatcttaatg aaggacttgg cagatgaact       24(
..
...
..
     acatgcctag tccaacattt tttcccagtg agtcacatcc tgggatccag tgtataaatc       150(
     caatatcatg tcttgtgcat aattcttcca aaggatctta ttttgtgaac tatatcagta       156(
     gtgtacatta ccatataatg taaaaagatc tacatacaaa caatgcaacc aactatccaa       162(
     gtgttatacc aactaaaacc cccaataaac cttgaacagt g                           166]
//
```

Fig. 5.6. Excerpt of a sample EMBL nucleotide sequence entry

5.3.1 Hidden treasures

Let's now return to the hidden treasures of nucleotide sequence databases. There are three categories of data useful to proteome studies and which complement the information found in SWISS-PROT and TrEMBL. These are unannotated protein sequences, new genes in old sequences and expressed sequence tags.

Unannotated protein sequences. It sometimes happens that authors who submit a nucleotide sequence forget to indicate which part of their DNA sequence codes for a protein sequence. This information is normally stored in the feature table under a key called 'CDS' (for CoDing Sequence). If such a feature is absent, the corresponding protein will not be integrated in TrEMBL and will probably also be missing in SWISS-PROT. It must be noted that the nucleotide databases are currently making an effort to detect these unannotated CDS and to update these entries. We therefore believe that these "forgotten" protein sequences will become rarer and rarer.

New genes in old sequences. There is a wealth of yet undiscovered genes and therefore proteins in existing DNA sequence entries. This is due to four different reasons which we will describe here. The first reason is quite mundane. In genomes of prokaryotes or fungi where gene density is high it often happens that a section of genomic DNA that contains a gene of interest will also contain an overlapping part of a flanking gene. These fragments will often fail to be detected and/or annotated. The second reason is due to sequencing errors that can introduce stop codons or frameshifts which can hide a protein coding region to the eyes (and computer programs) of the group that reported that sequence. The third reason is due to the failure to detect some genes whose intron/exon structure are quite complex. This is becoming an important issue now that large scale sequencing of genomic DNA from species such as the worm *Caenorhabditis elegans* and humans has started. It is more and more apparent that existing exon detecting algorithms are far from perfect. The programs using these algorithms will sometimes completely miss a gene, but are more likely to predict incorrect protein sequences where some regions are lacking or are added. This is a major problem for researchers working on proteome projects in eukaryotic organisms. We estimate that about 10 to 20% of the predicted proteins in *C. elegans* are currently incorrectly translated.

The last reason is due to a general bias against, or lack of confidence in the detection of small proteins. For example the yeast genome project applied as minimum size cutoff for potential coding regions of 300 base pairs, which corresponds to a protein of 100 residues. This cutoff was useful to exclude many artificial candidates from the list of potential proteins, as any DNA sequence will contain open reading frames whose size range can be computed according to the compositional bias of that sequence. However, it also excluded numerous real but small proteins. Exactly how many such proteins exist is difficult to estimate (as many of them are still unknown), but we can point out that in June 1997, there were 65 proteins of less than 100 residues biochemically or genetically characterised in the yeast genome, representing more than 1% of all yeast proteins.

It must be noted that methods have recently been developed to detect as yet unidentified genes in old sequences (Borodovsky et al. 1994, 1995), frameshift errors (Claverie 1993; Birney et al. 1996) and "tiny" genes (Barry et al. 1996). Large scale application of these methods should help solve some of the above problems.

Expressed sequence tags. A powerful approach to the exploration of the coding potential of a genome is the use of expressed sequence tags (ESTs) (Adams et al. 1991). ESTs are short sequences of 250 to 400 base pairs obtained by random, single-pass sequencing of cDNA libraries. Different sequencing strategies have been used, the most frequent including sequencing from the 5' end of the cDNA and 3' sequencing proximal to the poly(A) tail. Many eukaryotic genomes have been or currently are the target of large scale EST sequencing projects. In June 1997, the nucleotide sequence databases contained 1 million ESTs, 74% of them originating from human cDNA libraries. By definition ESTs are partial sequences and do not code (except in very rare cases) for complete protein sequences. As they are also very inaccurate due to the high level of sequence errors inherent to single-pass sequencing it would be logical to assume that they are not going to be very useful for proteome studies. However they can be useful when used in conjunction with high throughput partial sequencing using mass spectrometry. Programs have been developed (James et al. 1994; Gevaert et al. 1996) that scan all six potential frames of ESTs to find matches with partial protein sequences. If one or more matches are found it is sometimes feasible to reconstruct a significant part of a complete protein sequence from overlapping ESTs. This technique is so far only applicable to human proteins as it is the only species for which a very large quantity of ESTs is available. However it should be noted that while the number of ESTs in databases is very large, this does not necessarily mean that all proteins from, for example humans will be represented. Proteins produced from low abundance mRNAs may either not be present or difficult to find in EST databases. A last point about ESTs is that they are stored in international nucleotide sequence databases but they are also available in a specialised database, dbEST (Boguski et al. 1993) which offers additional annotations.

5.4 Pattern and profile databases

In some cases the sequence of an unknown protein is too distantly related to any protein of known primary structure to detect its resemblance by similarity searches. However relationships can be revealed by the presence of a particular cluster of residue types in the sequence, that are variously known as a pattern, motif, signature, or fingerprint. These motifs arise because specific region(s) of a protein which may be important, for example, for their binding properties or for their enzymatic

activity, are conserved in both structure and sequence. These structural requirements impose very tight constraints on the evolution of these small but important
portion(s) of a protein sequence. The use of protein sequence patterns or profiles to
determine the function of proteins is an essential tool of sequence analysis. This
reality was recognised more than 10 years ago:

*"There are many short sequences that are often (but not always) diagnostics of
certain binding properties or active sites. These can be set into a small subcollec-
tion and searched against your sequence"* (Doolittle 1986).

This observation spawned the development of databases that generally consist of
two components: a "discriminator" section which is used by a specifically designed
program to detect if a sequence belongs to a family or contains a specific domain,
and a "textual" section that provides a description of that protein family or domain.
An analogy can be drawn with the techniques used by police forces to identify a
criminal. Instead of a storing a full description of a particular individual, they store
an image of his fingerprint (the discriminator) along with a small file describing his
criminal history (the textual section).

5.4.1 PROSITE

PROSITE (Bairoch et al. 1997) is a database that currently describes over 1,000
protein families and domains. Most of these families or domains are detected using
one or more sequence patterns. Patterns are more formally known in the computer
sciences as regular expressions. Here is an example of pattern in PROSITE regular
expression notation:

$$K-[RG]-x(2)-C-\{DE\}-x(3,5)-[LIVM]$$

This pattern should be understood as follows: Lys; followed by either Arg or
Gly; two unspecified residues; a Cys; any residue but Asp or Glu; from three to
five unspecified residues; and finally either Leu, Ile, Val or Met.

PROSITE patterns are generated using the following procedure. One or more
"seed" patterns are manually defined using information derived from a multiple
alignment of the sequences belonging to the set of proteins under consideration.
When it is possible to do so, these patterns include residues and/or regions thought
or proved to be important to the biological function of that group of proteins.
Examples are active sites of enzymes, prosthetic group attachment sites, amino
acids involved in binding a metal ion, or cysteines involved in disulfide bonds. The
seed patterns are then used to scan SWISS-PROT to check that they are not matching with too many false positives (protein sequences that do not belong to the
group under consideration). If necessary the patterns are modified and re-scanned.
This iterative process continues until an optimal solution is obtained.

While sequence patterns are very useful, there are a number of protein families as
well as functional or structural domains that cannot be detected using patterns due
to their extreme sequence divergence. Typical examples are immunoglobulin, PH

or SH3 domains. In such domains there are only a few sequence positions which are well conserved. Any attempt to build a consensus pattern for such regions will either fail to pick up a significant proportion of the protein sequences that contain such a region (false negatives) or will pick up too many proteins that do not contain the region (false positives). The use of techniques based on profiles (also known as weight matrices) allows the detection of such proteins or domains. A profile is a table of numeric weights and gap costs that are position-specific for the amino acids in a sequence. These numbers are used to calculate a similarity score for any alignment between a profile and a sequence, or parts of a profile and a sequence. An alignment with a similarity score higher than or equal to a given cutoff value constitutes a motif occurrence. Another feature that distinguishes patterns from profiles is that the latter are usually not confined to small regions with high sequence similarity. Rather, they attempt to characterise a protein family or domain over its entire length.

As shown in Fig 5.7, PROSITE consists of two parts, each a separate file. The first file contains textual information that documents a protein family or domain and describes the patterns or profiles. This file can be considered as part of an on-line encyclopaedia of protein families. The second file in PROSITE contains all the necessary information for programs that scan sequence(s) for the occurrence of the patterns and/or profiles. This file also includes statistics on the number of hits obtained while scanning for that pattern or profile in SWISS-PROT as well as cross-references to the corresponding SWISS-PROT entries.

In addition to patterns and profiles that are targeted to the detection of specific protein families or domains, PROSITE also contains a number of patterns for the prediction of sites of amino acid post-translational modifications — for example N-glycosylation sites — as well as for the detection of topogenic sequences. Topogenic sequences are regions of a protein involved in the localisation or subcellular compartmentalisation of a protein product, for example, an endoplasmic reticulum targeting sequence.

5.4.2 BLOCKS

BLOCKS (Henikoff et al. 1997) is a database of... blocks! These are multiple sequence alignments containing no gaps and that correspond to the most highly conserved regions of proteins. These blocks are automatically generated by a two-step process: first by analysing the most highly conserved regions in groups of proteins represented in PROSITE; and then by calibrating these aligned protein segments against SWISS-PROT to obtain a measure of the chance distribution of matches. The final calibrated blocks make up the BLOCKS database. As BLOCKS is directly derived from PROSITE, it does not need specific documentation entries, but rather makes use of the PROSITE document entries. BLOCKS is associated with software tools that scan a sequence for the presence of these blocks. The results of such a scan are generally more sensitive than those produced by a regular

Part A: documentation section

```
{PDOC00263}
{PS00291; PRION_1}
{PS00706; PRION_2}
{BEGIN}
****************************
* Prion protein signatures *
****************************
```

Prion protein (PrP) [1,2,3] is a small glycoprotein found in high quantity in
the brains of humans or animals infected with a number of degenerative
neurological diseases such as Kuru, Creutzfeldt-Jacob disease (CJD), scrapie
or bovine spongiform encephalopathy (BSE).   PrP is encoded in the host genome
and expressed both in normal and infected cells. It has a tendency to
aggregate yielding polymers called rods.

Structurally, PrP is a protein consisting of a signal peptide, followed by
an N-terminal domain that contains tandem repeats of a short motif (PHGGGWGQ
in mammals, PHNPGY in chicken), itself followed by a highly conserved domain
of about 140 residues that contains a disulfide bond. Finally comes a C-
terminal hydrophobic domain post-translationally removed when PrP is attached
to the extracellular side of the cell membrane by a GPI-anchor.   The structure
of PrP is shown in the following schematic representation:

```
+---+----------------+-******--------------------****-----+-----+
|Sig| Tandem repeats |                      C       C   S|     |
+---+----------------+--------------------|--------|----|+-----+
                                          +--------+    |
                                                        GPI
```

'C': conserved cysteine involved in a disulfide bond.
'*': position of the patterns.

As signature pattern for PrP, we selected a perfectly conserved alanine- and
glycine-rich region of 16 residues as well as a region centered on the second
cysteine involved in the disulfide bond.

-Consensus pattern: A-G-A-A-A-A-G-A-V-V-G-G-L-G-G-Y
-Sequences known to belong to this class detected by the pattern: ALL.
-Other sequence(s) detected in SWISS-PROT: NONE.

-Consensus pattern: E-x-[ED]-x-K-[LIVM](2)-x-[KR]-[LIVM](2)-x-[QE]-M-C-x(2)-
                    Q-Y
                    [C is involved in a disulfide bond]
-Sequences known to belong to this class detected by the pattern: ALL.
-Other sequence(s) detected in SWISS-PROT: NONE.

-Last update: March 1997 / Text revised.

[1] Stahl N., Prusiner S.B.
     FASEB J. 5:2799-2807(1991).
[2] Brunori M., Chiara Silvestrini M., Pocchiari M.
     Trends Biochem. Sci. 13:309-313(1988).
[3] Prusiner S.B.
     Annu. Rev. Microbiol. 43:345-374(1989).
[E1] http://bioinformatics.weizmann.ac.il/hotmolecbase/prp.htm
{END}

Part B: pattern description

```
ID   PRION_1; PATTERN.
AC   PS00291;
DT   APR-1990 (CREATED); DEC-1992 (DATA UPDATE); NOV-1995 (INFO UPDATE).
DE   Prion protein signature 1.
PA   A-G-A-A-A-A-G-A-V-V-G-G-L-G-G-Y.
NR   /RELEASE=32,49340;
NR   /TOTAL=25(25); /POSITIVE=25(25); /UNKNOWN=0(0); /FALSE_POS=0(0);
NR   /FALSE_NEG=0; /PARTIAL=0;
CC   /TAXO-RANGE=??E??; /MAX-REPEAT=1;
DR   P10279, PRIO_BOVIN, T; Q01880, PRIP_BOVIN, T; P40242, PRIO_TRAST, T;
DR   P40243, PRP2_TRAST, T; P04156, PRIO_HUMAN, T; P40245, PRIO_AOTTR, T;
DR   P40246, PRIO_ATEGE, T; P40247, PRIO_CALJA, T; P40248, PRIO_CALMO, T;
DR   P40249, PRIO_CEBAP, T; P40250, PRIO_CERAE, T; P40251, PRIO_COLGU, T;
DR   P40252, PRIO_GORGO, T; P40253, PRIO_PANTR, T; P40254, PRIO_MACFA, T;
DR   P40255, PRIO_MANSP, T; P40256, PRIO_PONPY, T; P40257, PRIO_PREFR, T;
DR   P40258, PRIO_SAISC, T; P04273, PRIO_MESAU, T; P04925, PRIO_MOUSE, T;
DR   P40244, PRIO_MUSVI, T; P13852, PRIO_RAT  , T; P23907, PRIO_SHEEP, T;
DR   P27177, PRIO_CHICK, T;
DO   PDOC00263;
//
```

Fig. 5.7. A sample PROSITE entry

expression. Thus as a database *per se*, BLOCKS is not so useful, but as a method of identification of protein families or domains it is generally more powerful than PROSITE. However this is not true for PROSITE entries linked to profiles, these profiles being more sensitive than blocks.

5.4.3 PRINTS

PRINTS (Attwood et al. 1997) is a database of protein fingerprints. Each fingerprint is defined as a series of conserved local sequence alignments, here called motifs, which have been iteratively refined from database searches. The organisational principles and the aim of PRINTS are very close to those of PROSITE. The main difference lies in the pattern recognition methodology: instead of regular expression and profiles, PRINTS uses composite fingerprints made of multiple motifs. This technique offers, in some cases, a high level of discrimination. As shown in Fig. 5.8, PRINTS entries encapsulate both a description of the protein family that they aim to detect and the information on the fingerprints.

5.4.4 Pfam

Pfam (Eddy 1996) is a database of multiple alignments of protein domains or conserved protein regions. Each entry is based on a hand-edited seed alignment which contains representative protein sequences. A powerful statistical technique called a "hidden Markov model" (HMM) is then used to derive a descriptor from the seed alignment, which can be used to find the domain in a new protein or to realign a set of sequences to the model. Pfam distributes a full automatic alignment of all the sequences that the HMM descriptor has found to be members of the set in consideration. Each entry also comes with an annotation file which contains the parameters for these methods, a brief description of the domain and links to other databases.

5.4.5 Other protein domain databases

ProDom (Sonnhammer and Kahn 1994) is an automatically generated protein domain database. It is produced by an algorithm called DOMAINER that detects and clusters together homologous domains obtained from an exhaustive all-against-all similarity search in SWISS-PROT. It has been designed to assist with the analysis of the domain arrangement of proteins. ProDom provides a multiple sequence alignment for each of its domains. A major drawback of ProDom is a complete lack of annotation; each domain is solely associated with a serial number that changes from release to release. In fact ProDom should be considered more as a software tool for the display of the domain structure of a protein rather than a *bona fide* database.

```
LEPTIN View alignment              OBESITY FACTOR (LEPTIN) SIGNATURE
 Type of fingerprint: COMPOUND with 7  elements

 Creation date 22-MAR-1996

  1. ZHANG, Y., PROENCA, R., MAFFEI, M., BARONE, M., LEOPOLD, L. and
  FRIEDMAN, J.M.
  Positional cloning of the mouse obese gene and its human homologue.
  NATURE 372 425-432 (1994).

  Leptin, a metabolic monitor of food intake and energy need, is expressed
  by the ob obesity gene. The protein may function as part of a signalling
  pathway from adipose tissue that acts to regulate the size of the body
  fat depot [1], the hormone effectively turning the brain's appetite
  message off when it senses that the body is satiated. Obese humans have
  high levels of the protein, suggesting a similarity to type II (adult
  onset) diabetes, in which sufferers over-produce insulin, but can't respond
  to it metabolically - they have become insulin resistant. Similarly, it is
  thought that obese individuals may be leptin resistant.

  LEPTIN is a 7-element fingerprint that provides a signature for obesity
  factor proteins. The fingerprint was derived from an initial alignment of
  4 sequences: the motifs were drawn from well-conserved regions spanning
  the full alignment length. A single iteration on OWL27.0 was required to
  reach convergence, no further sequences being identified beyond the
  starting set.

  SUMMARY INFORMATION
       4 codes involving  7 elements
       0 codes involving  6 elements
       0 codes involving  5 elements
       0 codes involving  4 elements
       0 codes involving  3 elements
       0 codes involving  2 elements

  COMPOSITE FINGERPRINT INDEX

    7|   4    4    4    4    4    4    4
    6|   0    0    0    0    0    0    0
    5|   0    0    0    0    0    0    0
    4|   0    0    0    0    0    0    0
    3|   0    0    0    0    0    0    0
    2|   0    0    0    0    0    0    0
  --+-------------------------------------
     |   1    2    3    4    5    6    7

True positives..
 OB_BOVIN  OB_HUMAN  OB_MOUSE  OB_RAT


FINAL MOTIF SETS

  LEPTIN1              Length of motif = 19   Motif number = 1
  Leptin motif I - 1
                                   PCODE        ST      INT
  FLWLWSYLSYVQAVPIQKV              OB_MOUSE      9       9
  FLWLWSYLSYVQAVPIHKV              OB_RAT        9       9
  FLWLWPYLFYVQAVPIQKV              OB_HUMAN      9       9
  FLWLWPYLSYVEAVPIRKV              OB_BOVIN      9       9
 ..
 ...
 ..
  LEPTIN7             Length of motif = 16   Motif number = 7
  Leptin motif VII - 1
                                   PCODE        ST      INT
  LQGSLQDILQQLDVSP                 OB_MOUSE      150     0
  LQGSLQDILQQLDLSP                 OB_RAT        150     0
  LQGSLQDMLWQLDLSP                 OB_HUMAN      150     0
  LQGSLQDMLRQLDLSP                 OB_BOVIN      150     0
```

Fig. 5.8. Excerpt of a sample PRINTS entry

SBASE (Fabian et al. 1997) is a collection of protein domain sequences. The majority of these sequences are automatically extracted from SWISS-PROT entries using data from the feature tables. For example, if we consider part of the feature table of the SWISS-PROT entry for bovine enteropeptidase:

```
FT   DOMAIN       197    238    LDL-RECEPTOR CLASS A 1.
FT   DOMAIN       240    350    CUB.
FT   DOMAIN       358    520    MAM.
FT   DOMAIN       540    650    CUB.
FT   DOMAIN       657    695    LDL-RECEPTOR CLASS A 2.
FT   DOMAIN       694    787    SRCR.
```

SBASE will generate six separate entries from the above data that each contain the protein sequence segment corresponding to the annotated domain. SBASE can therefore be considered as a conversion of the protein sequence database into a series of entries corresponding to separate domains. The format of SBASE entries is based on that of SWISS-PROT.

5.4.6 Towards a unified database of protein families and domains

Recently, the groups that developed PROSITE, PRINTS and Pfam have decided to join forces to provide a comprehensive unified database of protein families and domains. The aim of this collaboration is to provide a single database of domain and family descriptions (mainly based on the current set of PROSITE documentation entries) linked to the different discriminator functions such as regular expressions, profiles, fingerprints and HMM which will be provided by the different groups. This will simplify the life of the user who will no longer need to "shop" in different places for a comprehensive domain analysis of new sequences.

5.5 2-D PAGE databases

While it is only recently that databases of proteins identified on two-dimensional polyacrylamide gel electrophoresis (2-D PAGE) have been developed and made available to the scientific community, they have already made a significant impact on proteome research. Furthermore, as described in Sect. 6.3.4, these databases are being used to pioneer methods of data integration across the WWW.

2-D PAGE databases contain two separate components:

1. Image data. This consists of one or more reference gel maps for a given biological sample. This sample is either a tissue, a physiological fluid or a cell in the case of free-living organisms. Two types of maps exist: those which are built from a single master 2-D gel and those which are derived from an assembled mosaic of 2-D gel masters where each covers a different section of the pH and molecular weight (MW) range. Currently the mosaic approach has not yet been

implemented in publicly available 2-D PAGE databases but it is expected that it will play a significant role in the near future. In both cases the image which is used as a reference map is generally a scanned representation of the gel, but occasionally a synthetic representation is used where all spots have been modelled using a bidimensional Gaussian function.

2. Textual information on each of the spots that has been identified on given master gels. In the majority of the 2-D PAGE databases this textual section includes data on the apparent molecular weight (Mr) and estimated isoelectric point (pI) of identified spots, the name of the protein, the method of identification, bibliographical references and cross-references to SWISS-PROT and other databases. In rare cases this section can also include amino acid composition and/or mass spectrometric data.

Access to the information stored in these databases is in most cases effected by a number of methods which include: clicking on a spot on a gel image, selecting from a list of identified proteins or by keyword searches.

5.5.1 SWISS-2DPAGE

SWISS-2DPAGE (Sanchez et al. 1995; Appel et al. 1996) is a database created and maintained at the University Hospital of Geneva in collaboration with the Department of Medical Biochemistry at Geneva University. It contains 2-D reference maps and information on identified proteins from a variety of human biological samples, namely: liver, kidney, plasma, HepG2, HepG2 secreted proteins, red blood cell, lymphoma, cerebrospinal fluid, macrophage-like cell line, erythroleukemia cell, astrocytes, colon and platelet, as well as for *S. cerevisiae*, *E. coli* and the slime mold *Dictyostelium discoideum*. Most proteins in SWISS-2DPAGE have been identified by the methods of microsequencing, immunoblotting, gel comparison or amino acid composition.

As shown in Fig. 5.9 (see p. 135), the format of the textual part of SWISS-2DPAGE is highly similar to that of SWISS-PROT. In addition both databases share entry names (on the ID line) and primary accession numbers (on the AC line). As with SWISS-PROT entries, SWISS-2DPAGE makes use of DE, OS, OC, reference lines, CC and DR lines. Three line types are specific to SWISS-2DPAGE:

1. The MT (MasTer) lines list the types of 2-D reference maps on which the protein has been found. This is particularly relevant to human proteins which can be identified on different maps corresponding to various biological samples.
2. The IM (IMages) lines list the 2-D PAGE images which are associated with the entry.
3. The 2D (2-Dimensional gel) lines give specific information such as mapping procedure (e.g. matching with another gel, microsequencing), the number, Mr and pI of spots, their amino acid composition, or describes normal and pathological variants.

SWISS-2DPAGE can be accessed in three ways. Firstly, by keyword search on the protein name, accession number or referenced author; secondly, from a SWISS-PROT entry, by selecting the hypertext link in the cross-reference (DR) line to SWISS-2DPAGE. If no corresponding entry exists in SWISS-2DPAGE, a virtual entry is created, from which one may view any of the 2-D reference maps. The theoretical pI and Mr for the protein are computed and its hypothetical position is shown on the map. Thirdly, by selecting a spot on one of the 2-D reference maps.

5.5.2 Other 2-D PAGE databases

There are an increasing number of 2-D PAGE databases. You will find an up-to-date list in WORLD-2DPAGE, an index to 2-D PAGE databases and services (see Table 5.1). However it must be noted that the scope of 2-D PAGE databases is not yet complete and that so far only a few species and human tissues have a dedicated 2-D PAGE database. The species that are currently represented are: yeast which benefits from two dedicated databases, YPM (Boucherie et al. 1996) from the Institut de Biochimie et Génétique Cellulaires in Bordeaux and YEAST 2D-PAGE at the University of Göteborg; *E. coli* with ECO2DBASE (VanBogelen et al. 1996); *Bacillus subtilis* with Sub2D (Antelmann et al. 1997); *Synechocystis* strain PCC 6803 with Cyano2Dbase (Sazuka and Ohara 1997); *Haemophilus influenzae* (Cash et al. 1997) and *Neisseria meningitidis* (Cash et al. 1995) from the University of Aberdeen; maize from the Institut National de la Recherche Agronomique (INRA); *Drosophila melanogaster* (Ericsson et al. 1997) at the Karolinska Institute; mouse from the Danish Centre for Human Genome Research and from Large Scale Biology (LSB); and rat liver also from LSB (Anderson et al. 1995).

For human tissues, there are many 2-D databases in addition to those mentioned in the description of SWISS-2DPAGE. These include: three databases for heart proteins, HSC-2DPAGE (Corbett et al. 1994) from Harefield Hospital, HEART-2DPAGE (Jungblut et al. 1994) from the German Heart Institute in Berlin and HP-2DPAGE from the MDC in Berlin; keratinocytes, MRC-5 fibroblasts, urine and bladder carcinomas from the Danish Centre for Human Genome Research (Celis et al. 1996); embryonal stem cells from the University of Edinburgh; A375 cell line from UCSF; breast cells (Giometti et al. 1997) from Argonne National Laboratory; and colon carcinoma (Ji et al. 1997) from the Melbourne Ludwig Institute for Cancer Research.

5.6 Three-dimensional structure databases

As explained in Chap. 7, insight into the three-dimensional (3-D) structure of a protein are crucial for the understanding of its precise function and are important for drug design and other biotechnological applications. Due to technical difficulties the elucidation of 3-D structures is a time consuming process. For a period of about 30 years the number of experimentally determined 3-D structures increased linearly up to a value of about 4,000 in late 1995. But in the last two years, the combination of new X-ray detectors, nuclear magnetic resonance (NMR) methods and the possibility offered by modern biotechnological methods to rapidly express and purify specific biologically interesting domains of complex proteins have suddenly increased the rate of appearance of new 3-D structures.

There is currently only a single database where 3-D structures are stored but there are also a small number of information resources that make direct use of this data.

5.6.1 PDB

The Protein Data Bank (PDB) (Abola et al. 1996) is an archive of 3-D structures of proteins, nucleic acids and other biological macromolecules. PDB, which is maintained by the Brookhaven National Laboratory, is currently undergoing major changes which will have an impact on the internal structure of the database as well as its interaction with its user community. The acronym PDB is also going to be replaced by "3DB" as the database will soon be known as the "Three-Dimensional Database of Biomolecular Structures".

In July 1997, there were 6,000 structures in PDB, 93% of which are proteins. However this number does not accurately represent the number of different proteins for which a 3-D structure is available. A significant number of proteins are present in PDB many times for a variety of reasons: either because they have been crystallised under different conditions or by different groups, or because they have been subjected to site-directed mutagenesis and different entries have been deposited for each of the variants. An extreme case is phage T4 lysozyme, a protein of 164 residues for which 177 different structures are available! There are in fact only about 1,700 different proteins for which a 3-D structure can be found in PDB.

The current format of PDB entries dates back to its conception more than 25 years ago. As mentioned above it will soon change. However, for a long period of time the old format will persist as many 3-D display and analysis software packages are dependent on this format. We show in Fig. 5.10 an excerpt of a PDB entry. As can be seen from this example, there are two main parts to each entry. Firstly, a header section that contains the name and biological source of the protein, bibliographical information, data concerning the experimental conditions used to obtain and refine the 3-D structure, cross-reference to SWISS-PROT, the amino acid sequence in three-letter code as well as some other data items. Secondly, the data concerning the spatial (X, Y, Z) coordinates of each of the atoms present in the structure. It is this information which is used by 3-D structure display programs such as RASMOL (Sayle and Milner-White 1995) or Swiss-PdbViewer (Guex and Peitsch 1996) (see Chap. 7; Fig. 7.1, p. 143).

```
HEADER    CYTOKINE                                    21-APR-95   1ILK      1ILK
COMPND    INTERLEUKIN-10                                                    1ILK
SOURCE    MOLECULE: INTERLEUKIN-10;                                         1ILK
SOURCE    2 ORGANISM_SCIENTIFIC: HOMO SAPIENS;                             1ILK
SOURCE    3 ORGANISM_COMMON: HUMAN;                                         1ILK
SOURCE    4 EXPRESSION_SYSTEM: ESCHERICHIA COLI                             1ILK
EXPDTA    X-RAY DIFFRACTION                                                 1ILK
AUTHOR    A.ZDANOV,C.SCHALK-HIHI,A.GUSTCHINA,A.WLODAWER                     1ILK
REVDAT    1   10-JUL-95 1ILK     0                                          1ILK
JRNL         AUTH   A.ZDANOV,C.SCHALK-HIHI,A.GUSTCHINA,M.TSANG,            1ILK
JRNL         AUTH 2 J.WEATHERBEE,A.WLODAWER                                 1ILK
JRNL         TITL   INTERLEUKIN-10 CRYSTAL STRUCTURE REVEALS THE           1ILK
JRNL         TITL 2 FUNCTIONAL DIMER WITH AN UNEXPECTED TOPOLOGICAL        1ILK
JRNL         TITL 3 SIMILARITY TO INTERFERON GAMMA                         1ILK
JRNL         REF    STRUCTURE  V. 3  591 1995                              1ILK
REMARK    1                                                                1ILK
REMARK    2 RESOLUTION. 1.8  ANGSTROMS.                                    1ILK
REMARK    3                                                                1ILK
REMARK    3 REFINEMENT.                                                     1ILK
REMARK    3    PROGRAM 1                     X-PLOR                         1ILK
REMARK    3    AUTHORS 1                     BRUNGER                        1ILK
REMARK    3    PROGRAM 2                     PROFFT                         1ILK
REMARK    3    AUTHORS 2                     KONNERT,HENDRICKSON,FINZEL     1ILK
REMARK    3    R VALUE                       0.156                         1ILK
REMARK    3    MEAN B VALUE                  22.4    ANGSTROMS**2           1ILK
REMARK    3    FINAL RMS COORD. SHIFT        0.007   ANGSTROMS             1ILK
REMARK    3    NUMBER OF REFLECTIONS         13105                         1ILK
REMARK    3    RESOLUTION RANGE       10.0 - 1.8   ANGSTROMS               1ILK
REMARK    3    DATA CUTOFF                   3.0     SIGMA(F)              1ILK
REMARK    3                                                                1ILK
REMARK    3 DATA COLLECTION.                                               1ILK
REMARK    3    NUMBER OF UNIQUE REFLECTIONS       15324                    1ILK
REMARK    3    COMPLETENESS OF DATA              80.8   %                  1ILK
REMARK    3    REJECTION CRITERIA               1.0    SIGMA(I)           1ILK
...
REMARK   18 EXPERIMENTAL DETAILS                                          1ILK
REMARK   18 DATE OF DATA COLLECTION           : SEP-94                    1ILK
REMARK   18 MONOCHROMATIC (Y/N)               : Y                         1ILK
REMARK   18 WAVELENGTH OR RANGE (A)           : 1.54                      1ILK
REMARK   18 DETECTOR TYPE                     : RAXIS-IIC                 1ILK
REMARK   18 DETECTOR MANUFACTURER             : SIEMENS                   1ILK
REMARK   18 INTENSITY-INTEGRATION SOFTWARE    : MSC PROGRAM PACKAGE       1ILK
...
REMARK  999 CROSS REFERENCE TO SEQUENCE DATABASE                          1ILK
REMARK  999 SWISS-PROT ENTRY NAME          PDB ENTRY CHAIN NAME           1ILK
REMARK  999    IL10_HUMAN                                                  1ILK
SEQRES    1    151 ASN SER CYS THR HIS PHE PRO GLY ASN LEU PRO ASN MET    1ILK
SEQRES    2    151 LEU ARG ASP LEU ARG ASP ALA PHE SER ARG VAL LYS THR    1ILK
SEQRES    3    151 PHE PHE GLN MET LYS ASP GLN LEU ASP ASN LEU LEU LEU    1ILK
SEQRES    4    151 LYS GLU SER LEU LEU GLU ASP PHE LYS GLY TYR LEU GLY    1ILK
SEQRES    5    151 CYS GLN ALA LEU SER GLU MET ILE GLN PHE TYR LEU GLU    1ILK
SEQRES    6    151 GLU VAL MET PRO GLN ALA GLU ASN GLN ASP PRO ASP ILE    1ILK
SEQRES    7    151 LYS ALA HIS VAL ASN SER LEU GLY GLU ASN LEU LYS THR    1ILK
SEQRES    8    151 LEU ARG LEU ARG LEU ARG ARG CYS HIS ARG PHE LEU PRO    1ILK
SEQRES    9    151 CYS GLU ASN LYS SER LYS ALA VAL GLU GLN VAL LYS ASN    1ILK
SEQRES   10    151 ALA PHE ASN LYS LEU GLN GLU LYS GLY ILE TYR LYS ALA    1ILK
SEQRES   11    151 MET SER GLU PHE ASP ILE PHE ILE ASN TYR ILE GLU ALA    1ILK
SEQRES   12    151 TYR MET THR MET LYS ILE ARG ASN                        1ILK
SSBOND    1 CYS     12    CYS    108                                       1ILK
SSBOND    2 CYS     62    CYS    114                                       1ILK
...
ATOM      1  N   ASN    10      13.720  42.109  57.534  1.00 72.51         1ILK
ATOM      2  CA  ASN    10      12.598  42.868  58.107  1.00 72.27         1ILK
ATOM      3  C   ASN    10      12.089  43.839  57.042  1.00 69.72         1ILK
ATOM      4  O   ASN    10      10.996  44.410  57.115  1.00 70.57         1ILK
ATOM      5  CB  ASN    10      12.752  43.358  59.535  1.00 79.07         1ILK
ATOM      6  CG  ASN    10      12.799  42.245  60.583  1.00 81.32         1ILK
ATOM      7  OD1 ASN    10      12.389  41.079  60.494  1.00 78.89         1ILK
ATOM      8  ND2 ASN    10      13.377  42.607  61.734  1.00 84.13         1ILK
...
```

Fig. 5.10. Excerpt of a sample PDB entry

5.6.2 Swiss-3DImage

Swiss-3DImage (Peitsch et al. 1995) is a database of high-quality annotated images of biological macromolecules with known 3-D structure. The goal of this image collection is to provide non-expert users with the essential structural information about any particular protein. The images are annotated to show key features such as active site residues, bound metal ions and disulfide bonds. The images are available in a variety of graphic file formats. A sample image showing the structure of insulin is shown in Fig. 5.11 (see p. 136).

5.6.3 DSSP, HSSP and FSSP

During the last 15 years, the group of Chris Sander at the EMBL and now at the EBI has developed a number of derived databases that merge structural information and sequence information. These databases are useful for the comparative studies of 3-D structures and to gain insight on the relationships between sequence, secondary structure elements and 3-D structure.

DSSP (Dictionary of Secondary Structure in Proteins) (Kabsch and Sander 1983) is a database that contains the derived information on the secondary structure and solvent accessibility for the protein 3-D structures stored in PDB.

HSSP (Homology-derived Secondary Structure of Proteins) (Schneider et al. 1997) is a database that consists of files containing alignments of the sequence of proteins with known 3-D structure with all their close homologs. As homologs are likely to have the same 3-D structure as the protein to which they have been aligned, this database is not only a database of aligned sequence families but also a database of implied secondary and tertiary structures. It should be noted that the proteins that are present in these alignments are likely to be amenable to the knowledge-based protein modelling methods described in Chap. 7.

FSSP (Families of Structurally Similar Proteins) (Holm and Sander 1997) is a database of structural alignments of proteins. It is based on an all-against-all comparison of the structures stored in PDB. Each entry in the database contains structural alignments of significantly similar proteins but excludes very close homologs — above 70% of sequence identity — because they are rarely markedly structurally different.

5.7 Post-translational modification databases

Most proteins are the target of post-translational modifications (PTM) and are not functional without being modified. The characterisation of PTMs is therefore an essential component of proteome projects (see Chap. 4 and 8). One would expect that many databases have been set up to serve this field. However this is currently

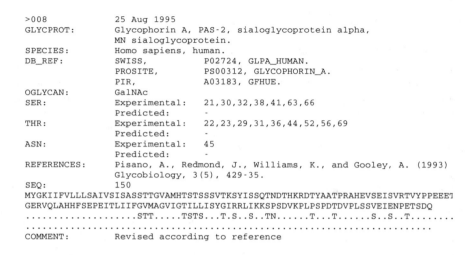

```
>008                25 Aug 1995
GLYCPROT:           Glycophorin A, PAS-2, sialoglycoprotein alpha,
                    MN sialoglycoprotein.
SPECIES:            Homo sapiens, human.
DB_REF:             SWISS,              P02724, GLPA_HUMAN.
                    PROSITE,            PS00312, GLYCOPHORIN_A.
                    PIR,                A03183, GFHUE.
OGLYCAN:            GalNAc
SER:                Experimental:   21,30,32,38,41,63,66
                    Predicted:      -
THR:                Experimental:   22,23,29,31,36,44,52,56,69
                    Predicted:      -
ASN:                Experimental:   45
                    Predicted:      -
REFERENCES:         Pisano, A., Redmond, J., Williams, K., and Gooley, A. (1993)
                    Glycobiology, 3(5), 429-35.
SEQ:                150
MYGKIIFVLLLSAIVSISASSTTGVAMHTSTSSSVTKSYISSQTNDTHKRDTYAATPRAHEVSEISVRTVYPPEEE1
GERVQLAHHFSEPEITLIIFGVMAGVIGTILLISYGIRRLIKKSPSDVKPLPSPDTDVPLSSVEIENPETSDQ
...................STT.....TSTS...T.S..S..TN......T...T......S..S..T........
............................................................................
COMMENT:            Revised according to reference
```

Fig. 5.12. A sample O-GLYCBASE entry

not the case and we are aware of only one publicly available PTM database (see below). One of the reasons for this is probably that SWISS-PROT contains a wealth of information on PTMs and the level of coverage of this information in that database is constantly increasing. A second reason is that there is not yet enough data either on the variations of the PTM structures themselves or the specific amino-acid sequences where they are found to justify the development of such databases.

5.7.1 O-GLYCBASE

O-GLYCBASE (Hansen et al. 1997) is a database of information on O-linked glycosylation sites in glycoproteins. As shown in Fig. 5.12, entries in O-GLYCBASE include information about species, sequence, glycosylation sites and glycan type and are cross-referenced to sequence databases.

5.8 Genomic databases

As was the case for nucleotide sequence databases, it is not immediately obvious why these databases should be mentioned in a book dealing exclusively with proteins. Genomic databases are a heterogeneous set of data resources that each deal with a specific biological organism. These databases contain a wealth of information on the genetic organisation of the species of interest. Information found in all these databases include: gene names, gene localisation (position on a chromosome) and cross-references to nucleotide and protein sequence databases. Some of these databases contain information useful for proteome studies, generally in the form of the description of the function of gene products or indirectly, by describing the effect that mutations of a specific gene have on the phenotype of the organism.

There are hundreds of genomic databases, we list here some of the best known examples: GDB — human (Fasman et al. 1997), MGD — mouse (Blake et al. 1997), AceDB — *C. elegans* (Durbin and Thierry-Mieg 1994), FlyBase — *D. melanogaster* (FlyBase consortium 1997), SGD — *S. cerevisiae*, MaizeDb — maize, AAtDB — *Arabidopsis thaliana* (Cherry et al. 1992), SubtiList — *B. subtilis* (Moszer et al. 1995), EcoGene (Rudd 1996) and ECD (Kroeger and Wahl 1997) — *E. coli*. These databases are also those where one is most likely to find information relevant to proteins.

The format of these databases is highly variable, but most of them are organised around relational database systems. You do not need to be knowledgeable about their detailed structure to use them as they generally offer WWW interfaces that can be queried to generate documents that are meant to be easy to read and understand. What one does need to know is that each of these databases is the product of a distinct scientific community which may have its idiosyncrasies and specific vocabulary (especially when it come to naming genes). For example, it helps to know that "fly people" tend to give quirky names to *Drosophila* genes, such as gang-of-three, godzilla, ghost, mother-against-madness, etc. They also do not see any harm in using the same acronym in lower or higher case to mean two separate things. For example: 'h' is the symbol of the 'hairy' gene while 'H' is the symbol of the 'hairless' gene!

In Fig. 5.13 we present a FlyBase entry for a gene coding for a lysozyme. Here, information relevant to the protein can be found in the sections with the header 'phenotypic information'.

As for specialised protein databases, you will find in Table 5.1 the address of a WWW document that offers a comprehensive list of genomic databases.

5.8.1 OMIM

MIM (Pearson et al. 1994) which stands for "Mendelian Inheritance in Man" was originally a printed catalogue of human genetic disorders authored and edited by Victor McKusick and his colleagues at Johns Hopkins. In recent years it evolved into a database of all known human genes, and is now known as "OMIM" (On-line MIM). It differs from a classical human genomic database such as GDB by the richness of textual information which is provided in each entry. Some of this information can be useful in the context of protein studies. One should, however, note that OMIM is developed and used by physicians, geneticists and other professionals concerned with genetic disorders and is therefore not specifically concerned with biochemical and sequence-oriented data. A further point is that the current

```
Gene symbol             LysP
Full name               Lysozyme P
FlyBase ID number       FBgn0004429
Secondary FlyBase id(s) FBgn0002575
Date                    23 May 97
Genetic location        3-[0]
Cytological location
   61F1--61F4
   Left limit from in situ hybridisation (FBrf0057272)
   Right limit from in situ hybridisation (FBrf0057272)
Prosite protein domain
   PS00128 == Alpha-lactalbumin / lysozyme C signature.
Enzyme name/number      lysozyme == EC 3.2.1.17
DNA/RNA accessions      X58383; g8200
Protein accessions      SWP/P29615  PIR/S20915
Size of product         141 aa
Transcript length       0.62
Genomic length of clone 32.5
RNA distribution        Salivary gland; adult.


Data from ref.  FBrf0057272 (Kylsten/1992)
   Cytological location    61F1--61F4 (determined by in situ
                           hybridisation)
   Size of product         141 aa
   Transcript length       0.62
   Genomic length of clone 32.5
   RNA distribution        Salivary gland; adult.
   Phenotypic information
      LysP has been cloned, sequenced and its developmental expression
      pattern analysed. The transcription levels of LysP decrease after
      bacterial injections into the haemocoel. The LysP gene product is
      therefore not induced as part of the immune response, but instead
      appears to have a role in the digestion of bacteria present in
      fermenting food.

Data from ref.  FBrf0068462 (Daffre/1994)
   Phenotypic information
      Lysozymes are expressed at a high level mainly in the digestive
      tract, they are not a component of the haemolymph. Lysozyme is
      not induced by bacteria so is not part of the inducible immune
      response but instead has been recruited for digestion of
      symbiotic bacteria in the stomach.

References

Primary
   Daffre et al.., 1994, Molec. gen. Genet. 242(2): 152--162 [FBrf0068462]
   Hultmark et al.., 1986, Cell 44: 429--438 [FBrf0043910]
   Kylsten, 1991, Thesis, Stockholm University, Sweden, 81pp [FBrf0064869]
   Kylsten et al.., 1992, Molec. gen. Genet. 232: 335--343 [FBrf0057272]

Review
   Hultmark, 1996, EXS 75: 87--102 [FBrf0089693]
```

Fig. 5.13. A sample FlyBase entry

textual nature of OMIM makes any type of automatic parsing of phenotypic information difficult. A sample OMIM entry is shown in Fig. 5.14. The size of an OMIM entry can range from that shown in the figure up to the size of a comprehensive review article.

*600021 MAD PROTEIN; MAD; MAX-BINDING PROTEIN

The MAD and MXI1 (600020) genes encode proteins that belong to a distinct
subfamily of MAX interacting proteins. The MAX protein (154950) specifically
interacts with the MYC protein family (190080) by forming heterodimers
mediated by their basic-helix-loop-helix-leucine zipper interaction domains.
Binding to MAX is essential for MYC transcription and transforming activity;
MYC homodimers are inactive. Both MAD and MXI1 bind MAX in vitro, forming a
sequence-specific DNA-binding complex similar to the MYC-MAX heterodimer. MAD
and MYC compete for binding to MAX. In addition, MAD acts as a
transcriptional repressor, while MYC appears to function as an activator.
MXI1 also appears to lack a transcriptional activation domain. Therefore,
MXI1 and MAD might antagonize MYC function and are candidate tumor suppressor
genes. Edelhoff et al.. (1994) mapped the human MAD and MXI1 genes to 2p13 and
10q25, respectively, by fluorescence in situ hybridization. The homologous
gene in the mouse was mapped to chromosome 6 by interspecific backcross
analysis. Shapiro et al.. (1994) confirmed the assignments of the MAD and MXI1
genes to chromosomes 2p13-p12 and 10q24-q25, respectively, by somatic cell
mapping and fluorescence in situ hybridization.

The designation MAD was derived from MAX dimerizer (Eisenman, 1994).

FIELD RF
1. Edelhoff, S.; Ayer, D. E.; Zervos, A. S.; Steingrimsson, E.; Jenkins,
N. A.; Copeland, N. G.; Eisenman, R. N.; Brent, R.; Disteche, C. M.
: Mapping of two genes encoding members of a distinct subfamily of
MAX interacting proteins: MAD to human chromosome 2 and mouse chromosome
6, and MXI1 to chromosome 10 and mouse chromosome 19. Oncogene 9:
665-668, 1994.

2. Eisenman, R. N.: Personal Communication. Seattle, Wash. 7/27/1994.

3. Shapiro, D. N.; Valentine, V.; Eagle, L.; Yin, X.; Morris, S. W.;
Prochownik, E. V.: Assignment of the human MAD and MXI1 genes to
chromosomes 2p12-p13 and 10q24-q25. Genomics 23: 282-285, 1994.

FIELD CD
Victor A. McKusick: 7/1/1994
FIELD ED
mark: 08/27/1996

Fig. 5.14. A sample OMIM entry

5.9 Metabolic databases

Metabolic databases (Karp 1997) are a heterogeneous category of data resources
targeted toward the description of enzymes, biochemical reactions and metabolic
pathways. Some of these databases provide detailed descriptions of all known
enzymatic reactions catalysed by a specific organism while others tend to special-
ise in a subset of the biochemical pathways but will cater for a variety of organisms. In general these databases are tightly coupled with query software that allows
the user to visualise reaction schemes. From a point of view of their interconnec-
tion with other sources of data they differ from other biological databases in that
they are not only cross-referenced to sequence databases such as SWISS-PROT or
EMBL/Genbank but that they sometimes offer links to information resources on

```
ID   1.3.3.3
DE   COPROPORPHYRINOGEN OXIDASE.
AN   COPROPORPHYRINOGENASE.
AN   COPROPORPHYRINOGEN-III OXIDASE.
AN   COPROGEN OXIDASE.
CA   COPROPORPHYRINOGEN-III + O(2) = PROTOPORPHYRINOGEN-IX + 2 CO(2).
CF   IRON.
DI   COPROPORPHYRIA; MIM:121300.
PR   PROSITE; PDOC00783;
DR   P36553, HEM6_ECOLI;  P36551, HEM6_HUMAN;  P36552, HEM6_MOUSE;
DR   P43898, HEM6_PSEAE;  P33771, HEM6_SALTY;  P35055, HEM6_SOYBN;
DR   P11353, HEM6_YEAST;
//
```

Fig. 5.15. A sample ENZYME entry

chemical compounds. Metabolic databases are useful to a large gamut of research tasks, most notably when studying whole genomes to validate or predict the existence of specific enzymatic pathways. They are also used for the design or the modification of metabolic or catabolic pathways in a biotechnological context, for phylogenetic studies of molecular evolution as well as for computer-aided instruction.

Here, we will describe some representative metabolic databases, but before that we might note that the "ancestor" of these databases is the well known Boehringer Mannheim chart of metabolic pathways (Michal 1982) which still has its place on the wall of many laboratories. A version of this chart is now available on the WWW (see Table 5.1) as a series of linked and clickable images.

5.9.1 Some specific metabolic databases

ENZYME. The ENZYME database (Bairoch 1996) is not a metabolic database in the strictest sense as it is a repository of information relative to the nomenclature of enzymes. It is primarily based on the recommendations of the Nomenclature Committee of the International Union of Biochemistry and Molecular Biology (IUBMB) (NC-IUBMB 1992). It contains the following data for each type of characterised enzyme for which an EC (Enzyme Commission) number has been provided: the recommended name, alternative names (if any), the catalytic activity and cofactors (if any). It also includes cross-references to SWISS-PROT, PROSITE and OMIM. As shown in Fig. 5.15, the format of ENZYME is based on that of the 'SWISS-line' of databases (i.e. SWISS-PROT, PROSITE, SWISS-2DPAGE). ENZYME is used as the nomenclature source for enzyme names and reactions by most metabolic databases as well as by other biomolecular databases.

EcoCyc. EcoCyc (Karp et al. 1997) is a database of the intermediate metabolism of *E. coli* which currently describes 130 different pathways. It contains information on the enzymes that carry out each reaction, including their cofactors, activators, inhibitors and their subunit structures. EcoCyc contains a dictionary of chemical compounds that lists synonyms and chemical structures for each of the compounds involved in a reaction described in the database. The data structure of EcoCyc is quite complex but users generally access the database through an easy to use graphical interface. This interface can automatically generate drawings of metabolic pathways, as shown in Fig. 5.16 (see p. 137). EcoCyc is linked to ENZYME, SWISS-PROT, EMBL/Genbank, EcoGene and PDB. Recently the group responsible for the development of EcoCyc has developed a similar database, HinCyc, describing the metabolism of *H. influenzae*.

KEGG. KEGG (Kanehisa 1996), the Kyoto Encyclopaedia of Genes and Genomes, is an ambitious effort to computerise the current knowledge of molecular and cellular biology in terms of the information pathways that consist of interacting molecules or genes. KEGG consists of four types of data: pathway maps, molecule tables, gene tables and genome maps. The pathway maps, which are graphical diagrams, currently consist of a series of metabolic pathways not specific to any particular organism. By its scope, KEGG is not *stricto senso* a metabolic database as it is much more ambitious in its goals.

EMP/WIT. The Enzyme and Metabolic Pathways database (EMP) (Selkov et al. 1997) is the impressive endeavour of the group of Selkov in Russia which, for more than 10 years, has extracted information from the literature on metabolic pathways, enzyme activities, cofactors, kinetics, purification conditions, etc. EMP contains about 2,000 species-specific metabolic pathways with embedded links to chemical compounds and enzyme descriptions. EMP is a commercial database, but a subset of it is publicly accessible under the name WIT (What Is There). WIT provides text-based diagrams of pathways. It also provides tools to predict the existence of metabolic pathways in organisms whose complete genomes are available.

5.10 Conclusions

5.10.1 How to access all of this data

In this chapter we have seen only a small, representative subset of the databases that are available to life scientists working with proteins. If you have managed to read this chapter up to this point, you probably have a fair idea of what type of information is available. What is maybe less obvious is how one can access all of these information resources. A few years ago the answer would have been specific to each database, but fortunately there is now a simpler solution, the World-Wide Web (described in Sect. 6.3.2). Almost all of the resources mentioned in this chapter are available on the WWW and all are currently free of charge. Table 5.1 lists the WWW addresses of all these databases. Alternatively, if you prefer to make local copies of some databases on your computer system, almost all the organisations that develop these databases offer anonymous FTP (File Transfer Protocol) site(s) which allow you to download the relevant file(s). Links to these FTP sites are available from the corresponding WWW pages. You should however remember that while it can be useful to have a local copy of a database, it has a number of implications. Firstly, you need to make sure that you update your data set regularly. Optimally you should follow the specific update cycles of the databases that you have locally imported. The update frequency of databases can range from a week (e.g. SWISS-PROT) up to a year. Secondly, for databases with a complex data structure, such as it is generally the case for genomic or metabolic databases, you may need to rebuild the relational scheme used or implied by that particular database. This may be an involved and time-consuming task.

5.10.2 What future for protein databases?

Trying to extrapolate what will be the long term future of protein information resources is a dangerous exercise. Who would have predicted five years ago the emergence of the WWW and its tremendous impact on all aspects of information retrieval and processing. Nevertheless, some clear trends can be distinguished and some predictions can be made.

1. The amount of information available will continue to increase exponentially for quite a long time. Even when the major genome projects become complete, it is expected that there will be a strong pressure to continue to sequence genomes, both for the purpose of biotechnological applications and for the exploration of biodiversity. It is probable that in 2010 there will be thousands of complete genomes in databases. We also believe that a significant effort will be mounted to systematically study human polymorphism at the sequence level.

2. Proteome projects, both in academic and industrial environments, will soon provide a flood of data on the characterisation and functionalisation of the genome-derived sequence data. New databases will need to be set up to cater for specific types of information that will be obtained by a variety of analytical methods. Mass spectrometric methods will generate massive amounts of data on post-translational modifications. This trend can be conceptualised as the protein equivalent of the DNA expressed sequence tags (ESTs) in that it will provide a high-throughput, automated, rapid survey of proteomes.

3. New types of databases will be developed that deal with the interactions between proteins. These databases will be necessary to mount efforts to attempt to model biological processes within cells, as well as between cells. GIF-DB (Gene Interactions in the Fly DataBase) (Jacq et al. 1997) is the first example of this new category of databases. It describes molecular interactions involved in the process of embryonic pattern formation in *D. melanogaster*. KEGG (see Sect. 5.9.1) is another database which may evolve towards such an ambitious goal.

4. The distinction between databases and journals will become less obvious as increasing amount of data are published only in databases rather than in classical printed reports. This trend is already apparent for genomic data. It often happens that a single paper reports the sequencing of a complete genome. Yet thousands of protein sequence entries are created using that data, using information not mentioned in such an article. We believe that this trend will also be true for proteomic data.

5. It will become more and more difficult for single groups to maintain centralised large-scale databases. It is therefore expected that many databases will be the products of the collaboration of professional database developers interacting with a large number of expert users all over the world.

6. As it will be seen in the next chapter, integration of databases is of paramount importance for their long-term usefulness as well as for the development of higher level interrogation tools capable of intelligently retrieving data across a large gamut of heterogeneous data sets. We firmly believe that the future of protein databases is therefore dependent on the level and extent of their integration.

There is a famous Chinese aphorism which can be translated as "may you live in interesting times". It was originally used as a curse! But scientists will generally take it as an expression of hope that their field of research will bring them daily excitement and interesting vistas. The field of proteome research is without doubt going to experience some very interesting times and databases are going to be a major tool for the understanding of life processes.

Table 5.1. WWW addresses for the databases mentioned in Chap. 5

- An updated version of this table is available from:
  http://www.expasy.ch/users/springer97_table51.html
- The following list contains almost exclusively pointers to information sources for molecular biologists:
  http://www.expasy.ch/amos_www_links.html
- In rare cases, when no WWW site is available, an FTP address is provided
- Some databases are available on more than one server (these are called "mirrors"). Mirror sites are listed on separate lines

Protein sequence databases (Sect. 5.2)

SWISS-PROT	http://www.expasy.ch/sprot/sprot-top.html
TrEMBL	http://www.expasy.ch/srs5/
OWL	http://www.biochem.ucl.ac.uk/bsm/dbbrowser/OWL/OWL.html
Specialised protein db list	http://www.expasy.ch/amos_www_links.html#Gene_prot
YPD	http://quest7.proteome.com/YPDhome.html
GCRDb	http://receptor.mgh.harvard.edu/GCRDBHOME.html
AMSdb	http://www.univ.trieste.it/~nirdbbcm/tossi/pag1.html

Table 5.1. (cont.) WWW addresses for the databases mentioned in Chap. 5

Nucleotide sequence databases (Sect. 5.3)

EMBL	http://www.ebi.ac.uk/ebi_docs/embl_db/ebi/topembl.html
GenBank	http://www.ncbi.nlm.nih.gov/Web/Search/index.html
DDBJ	http://www.ddbj.nig.ac.jp/
dbEST	http://www.ncbi.nlm.nih.gov/dbEST/
	http://www.ebi.ac.uk/dbest/dbest_index.html

Patterns and profiles databases (Sect. 5.4)

PROSITE	http://www.expasy.ch/sprot/prosite.html
BLOCKS	http://www.blocks.fhcrc.org/
PRINTS	http://www.biochem.ucl.ac.uk/bsm/dbbrowser/PRINTS/PRINTS.html
Pfam	http://genome.wustl.edu/eddy/Pfam/Welcome.html
ProDom	http://protein.toulouse.inra.fr/
SBASE	http://base.icgeb.trieste.it/sbase/

2-D PAGE databases (Sect. 5.5)

WORLD-2DPAGE	http://www.expasy.ch/ch2d/2d-index.html
SWISS-2DPAGE	http://www.expasy.ch/ch2d/ch2d-top.html
YPM	http://www.ibgc.u-bordeaux2.fr/YPM
YEAST 2D-PAGE	http://yeast-2dpage.gmm.gu.se/
ECO2DBASE	http://pcsf.brcf.med.umich.edu/eco2dbase
Sub2D	http://pc13mi.biologie.uni-greifswald.de/
Cyano2Dbase	http://www.kazusa.or.jp/tech/sazuka/cyano/proteome.html
Aberdeen 2-D db	http://www.abdn.ac.uk/~mmb023/2dhome.htm
Maize 2-D db	http://moulon.moulon.inra.fr/imgd/
Fly 2-D db	http://tyr.cmb.ki.se/
LSB 2-D db	http://www.lsbc.com/2dmaps/patterns.htm
HSC-2DPAGE	http://www.harefield.nthames.nhs.uk/nhli/protein/
HEART-2DPAGE	http://www.chemie.fu-berlin.de/user/pleiss/dhzb.html
HP-2DPAGE	http://www.mdc-berlin.de/~emu/heart/
Danish 2-D db	http://biobase.dk/cgi-bin/celis
Emb. stem cell	http://www.ed.ac.uk/~nh/2DPAGE.html
A375 UCSF 2-D	http://rafael.ucsf.edu/2DPAGEhome.html
ANL Breast 2-D	http://www.anl.gov/CMB/PMG/projects/index_hbreast.html
Colon carcinoma	http://www.ludwig.edu.au/www/jpsl/jpslhome.html

Three-dimensional structure databases (Sect. 5.6)

PDB	http://www.pdb.bnl.gov
	http://www2.ebi.ac.uk/pdb/
Swiss-3DImage	http://www.expasy.ch/sw3d/sw3d-top.html
DSSP	ftp://ftp.ebi.ac.uk/pub/databases/dssp/
HSSP	ftp://ftp.ebi.ac.uk/pub/databases/hssp/
FSSP	ftp://ftp.ebi.ac.uk/pub/databases/fssp/

Post-translational modification databases (Sect. 5.7)

O-GLYCBASE	http://www.cbs.dtu.dk/OGLYCBASE/cbsoglycbase.html

Table 5.1. (cont.) WWW addresses for the databases mentioned in Chap. 5

Genomic databases (Sect. 5.8)

Genomic db list	http://www.expasy.ch/amos_www_links.html#Organisms
GDB	http://gdbwww.gdb.org/
MGD	http://www.informatics.jax.org/mgd.html
AceDB	http://moulon.inra.fr/acedb/acedb.html
FlyBase	http://flybase.bio.indiana.edu/
	http://www.ebi.ac.uk:7081/
SGD	http://genome-www.stanford.edu/Saccharomyces/
MaizeDb	http://www.agron.missouri.edu/
AAtDB	http://genome-www.stanford.edu/Arabidopsis/
SubtiList	http://www.pasteur.fr/Bio/SubtiList.html
EcoGene	Will soon be available on WWW
ECDC	http://susi.bio.uni-giessen.de/usr/local/www/html/ecdc.html
OMIM	http://www3.ncbi.nlm.nih.gov/omim/

Metabolic databases (Sect. 5.9)

Boehringer chart	http://www.expasy.ch/cgi-bin/search-biochem-index
ENZYME	http://www.expasy.ch/sprot/enzyme.html
EcoCyc	http://www.ai.sri.com/ecocyc/ecocyc.html
HinCyc	http://www.ai.sri.com/ecocyc/hincyc.html
KEGG	http://www.genome.ad.jp/kegg/
WIT	http://www.cme.msu.edu/WIT/

Conclusions (Sect. 5.10)

GIF-DB	http://gifts.univ-mrs.fr/GIF_DB/GIF_DB_home_page.html

References

Abola EE, Manning NO, Prilusky J, Stampf DR, Sussman JL (1996) The Protein Data Bank: current status and future challenges. J Res Natl Inst Stand Technol 101:231–241

Adams MD, Kelley JM, Gocayne JD, Dubnick M, Polymeropoulos MH, Xiao H, Merril CR, Wu A, Olde B, Moreno RF, Kerlavage AR, McCombie WR, Venter JC (1991) Complementary DNA sequencing: expressed sequence tags and human genome project. Science 252:1651–1656

Altschul SF, Gish W, Miller W, Myers EW, Lipman DJ (1990) Basic local alignment search tool. J Mol Biol 215:403–410

Anderson NL, Esquer-Blasco R, Hofmann JP, Meheus L, Raymackers J, Steiner S, Witzmann F, Anderson NG (1995) An updated two-dimensional gel database of rat liver proteins useful in gene regulation and drug effects studies. Electrophoresis 16:1977–1981

Antelmann H, Bernhardt J, Schmid R, Mach H, Volker U, Hecker M (1997) First steps from two-dimensional protein index towards a response regulation map for *Bacillus subtilis*. Electrophoresis 18, in press

Appel RD, Bairoch A, Hochstrasser DF (1994) A new generation of information retrieval tools for biologists: the example of the ExPASy WWW server. Trends Biochem Sci 19:258–260

Appel RD, Sanchez JC, Bairoch A, Golaz O, Ravier F, Pasquali C, Hughes GJ, Hochstrasser DF (1996) The SWISS-2DPAGE database of two-dimensional polyacrylamide gel electrophoresis, its status in 1995. Nucleic Acids Res 24:180–181

Apweiler R, Gateau A, Contrino S, Martin MJ, Junker V, O'Donovan C, Lang F, Mitaritonna N, Kappus S, Bairoch A (1997) Protein sequence annotation in the genome era: the annotation concept of SWISS-PROT + TREMBL. In: ISMB-97; Proceedings 5nd International Conference on Intelligent Systems for Molecular Biology, AAAI Press, Menlo Park

Attwood TK, Beck ME, Bleasby AJ, Degtyarenko K, Michie AD, Parry-Smith DJ (1997) Novel developments with the PRINTS protein fingerprint database. Nucleic Acids Res 25:212–216

Bairoch A (1996) The ENZYME data bank in 1995. Nucleic Acids Res 24:221–222

Bairoch A, Apweiler R (1997) The SWISS-PROT protein sequence data bank and its supplement TrEMBL. Nucleic Acids Res 25:31–36

Bairoch A, Bucher P, Hofmann K (1997) The PROSITE database, its status in 1997. Nucleic Acids Res 25:217–221

Barry C, Fichant G, Kalogeropoulos A, Quentin Y (1996) A computer filtering method to drive out tiny genes from the yeast genome. Yeast 12:1163–1178

Benson DA, Boguski MS, Lipman DJ, Ostell J (1997) GenBank. Nucleic Acids Res 25:1–6

Birney E, Thompson JD, Gibson TJ (1996) PairWise and SearchWise: finding the optimal alignment in a simultaneous comparison of a protein profile against all DNA translation frames. Nucleic Acids Res 24:2730–2739

Blake JA, Richardson JE, Davisson MT, Eppig JT (1997) The Mouse Genome Database (MGD). A comprehensive public resource of genetic, phenotypic and genomic data. Nucleic Acids Res 25:85–91

Bleasby AJ, Akrigg D, Attwood TK (1994) OWL — a non-redundant composite protein sequence database. Nucleic Acids Res 22:3574–3577

Boguski MS, Lowe TMJ, Tolstoshev CM (1993) dbEST — database for expressed sequence tags. Nat Genet 4:332–333

Borodovsky MY, Koonin EV, Rudd KE (1994) New genes in old sequence: a strategy for finding genes in the bacterial genome. Trends Biochem Sci 19:309–313

Borodovsky MY, McIninch JD, Koonin EV, Rudd KE, Medigue C, Danchin A (1995) Detection of new genes in a bacterial genome using Markov models for three gene classes. Nucleic Acids Res 23:3554–3562

Boucherie H, Sagliocco F, Joubert R, Maillet I, Labarre J, Perrot M (1996) Two-dimensional gel protein database of *Saccharomyces cerevisiae*. Electrophoresis 17:1683–1699

Cash P, Argo E, Abadi FJ (1995) Development of a 2-dimensional protein database for *Neisseria meningitidis*. International Congress of Electrophoresis, Paris

Cash P, Argo E, Langford P, Krol SJ (1997) Development of an *Haemophilus* 2D protein database. Electrophoresis 18, in press

Celis JE, Gromov P, Ostergaard M, Madsen P, Honoré B, Dejgaard K, Olsen E, Vorum H, Kristensen DB, Gromova I, Haunso A, Van Damme J, Puype M, Vandekerckhove J, Rasmussen HH (1996) Human 2-D PAGE databases for proteome analysis in health and disease: http://biobase.dk/cgi-bin/celis. FEBS Lett 398:129–134

Cherry JM, Cartinhour SW, Goodman HM (1992) AAtDB, an *Arabidopsis thaliana* database. Plant Mol Biol Rep 10:308–309

Claverie JM (1993) Detecting frame shifts by amino acid sequence comparison. J Mol Biol 234:1140–1157

Corbett JM, Wheeler CH, Baker CS, Yacoub MH, Dunn MJ (1994) The human myocardial two-dimensional gel protein database: update 1994. Electrophoresis 15:1459–1465

Doolittle RF (1986) Of URFs and ORFs: a primer on how to analyze derived amino acid sequences. University Science Books, Mill Valley, California

Durbin R, Thierry-Mieg J (1994) The ACEDB genome database. In: Suhai S (ed) Computational methods in genome research. Plenum Press, New York, pp 45–55

Eddy SR (1996) Hidden Markov models. Curr Opin Struct Biol 6:361–365

Ericsson C, Pethö Z, Mehlin H (1997) An on-line two-dimensional polyacrylamide gel protein database of adult *Drosophila melanogaster*. Electrophoresis 18:484–490

Fabian P, Murvai J, Hatsagi Z, Vlahovicek K, Hegyi H, Pongor S (1997) The SBASE protein domain library, release 5.0: a collection of annotated protein sequence segments. Nucleic Acids Res 25:240–243

Fasman KH, Letovsky SI, Li P, Cottingham RW, Kingsbury DT (1997) The GDB human genome database Anno 1997. Nucleic Acids Res 25:72–80

FlyBase Consortium (1997) FlyBase: a *Drosophila* database. Nucleic Acids Res 25:63–66

Gevaert K, Verschelde JL, Puype M, Van Damme J, Goethals M, De Boeck S, Vandekerckhove J (1996) Structural analysis and identification of gel-purified proteins, available in the femtomole range, using a novel computer program for peptide sequence assignment, by matrix-assisted laser desorption ionization-reflection time-of-flight-mass spectrometry. Electrophoresis 17:918–924

Giometti CS, Williams K, Tollaksen SL (1997) A two-dimensional electrophoresis database of human breast epithelial cell proteins. Electrophoresis 18:573–581

Guex N, Peitsch MC (1996) Swiss-PdbViewer: a fast and easy-to-use PDB viewer for Macintosh and PC. PDB Quat Newslett 77:7

Hansen JE, Lund O, Rapacki K, Brunak S (1997) O-GLYCBASE version 2.0: a revised database of O-glycosylated proteins. Nucleic Acids Res 25:278–282

Henikoff JG, Pietrokovski S, Henikoff S (1997) Recent enhancements to the Blocks database servers. Nucleic Acids Res 25:222–225

Holm L, Sander C (1997) Dali/FSSP classification of three-dimensional protein folds. Nucleic Acids Res 25:231–234

Jacq B, Horn F, Janody F, Gompel N, Serralbo O, Mohr E, Leroy C, Bellon B, Fasano L, Laurenti P, Röder L (1997) GIF-DB, a WWW database on gene interactions involved in *Drosophila melanogaster* development. Nucleic Acids Res 25:67–71

James P, Quadroni M, Carafoli E, Gonnet G (1994) Protein identification in DNA databases by peptide mass fingerprinting. Protein Sci 3:1347–1350

Ji H, Reid GE, Moritz RL, Eddes JS, Burgess AW, Simpson RJ (1997) A two-dimensional gel database of human colon carcinoma proteins. Electrophoresis 18:605–613

Jungblut P, Otto A, Zeindl-Eberhardt E, Pleissner KP, Knecht M, Regitz-Zagrosek V, Fleck E, Wittmann-Liebold B (1994) Protein composition of the human heart: the construction of a myocardial two-dimensional electrophoresis database. Electrophoresis 15:685–707

Kabsch W, Sander C (1983) Dictionary of protein secondary structure: pattern recognition of hydrogen-bonded and geometrical features. Biopolymers 22:2577–2637

Kanehisa M (1996) Toward pathway engineering: a new database of genetic and molecular pathways. Sci Tech Japan 59:34–38

Karp PD (1997) Metabolic databases. Trends Biochem Sci, in press

Karp PD, Riley M, Paley SM, Pelligrini-Toole A, Krummenacker M (1997) EcoCyc: enyclopedia of *Escherichia coli* genes and metabolism. Nucleic Acids Res 25:43–50

Kolakowski LF Jr (1994) A G-protein-coupled receptor database. Recept Channels 2:1–7

Kroeger M, Wahl R (1997) Compilation of DNA sequences of *Escherichia coli* K12: description of the interactive databases ECD and ECDC (update 1996). Nucleic Acids Res 25:39–42

Michal G (1982) Biochemical Pathways Wall Chart. Boehringer Mannheim GmbH Biochemica

Moszer I, Glaser P, Danchin A (1995) SubtiList: a relational data base for the *Bacillus subtilis* genome. Microbiology 141:261–268

NC-IUBMB (1992). Enzyme Nomenclature, Recommendations of the Nomenclature Committe of the International Union of Biochemistry and Molecular Biology on the Nomenclature and Classification of Enzymes. Academic Press, New-York

Payne WE, Garrels JI (1997) Yeast Protein Database (YPD): a database for the complete proteome *of Saccharomyces cerevisiae*. Nucleic Acids Res 25:57–62

Pearson PL, Francomano C, Foster P, Bocchini C, Li P, McKusick VA (1994) The status of online Mendelian inheritance in man (OMIM) medio 1994. Nucleic Acids Res 22:3470–3473

Peitsch MC, Wells TN, Stampf DR, Sussman JL (1995) The Swiss-3DImage collection and PDB-Browser on the World-Wide Web. Trends Biochem Sci 20:82–83

Robinson C (1994) The European Bioinformatics Institute (EBI) — open for business. Trends Biotechnol 12:391–392

Rudd KE (1996) *Escherichia coli* K-12 on the Internet. Trends Genet 12:156–157

Sanchez JC, Appel RD, Golaz O, Pasquali C, Ravier F, Bairoch A, Hochstrasser DF (1995) Inside SWISS-2DPAGE database. Electrophoresis 16:1131–1151

Sayle RA, Milner-White EJ (1995) RASMOL: biomolecular graphics for all. Trends Biochem Sci 19 :258–260

Sazuka T, Ohara O (1997) Towards a proteome project of cyanobacterium *Synechocystis* sp. strain PCC6803: linking 130 protein spots with their respective genes. Electrophoresis 18, in press

Schneider R, de Davuvar A, Sander C (1997) The HSSP database of protein structure-sequence alignments. Nucleic Acids Res 25:226–230

Selkov E, Galimova M, Goryanin I, Gretchkin Y, Ivanova N, Komarov Y, Maltsev N, Mikhailova N, Nenashev V, Overbeek R, Panyushkina E, Pronevitch L, Selkov E Jr (1997) The metabolic pathway collection: an update. Nucleic Acids Res 25:37–38

Sonnhammer EL, Kahn D (1994) Modular arrangement of proteins as inferred from analysis of homology. Protein Sci 3:482–492

Stoesser G, Sterk P, Tuli MA, Stoehr PJ, Cameron GN (1997) The EMBL Nucleotide Sequence Database. Nucleic Acids Res 25:7–13

Tateno Y, Gojobori T (1997) DNA Data Bank of Japan in the age of information biology. Nucleic Acids Res 25:14–17

VanBogelen RA, Abshire KZ, Pertsemlidis A, Clark RL, Neidhardt FC (1996) Gene-protein database of *Escherichia coli* K-12, edition 6. In: Neidhardt FC (ed) *Escherichia coli* and *Salmonella typhimurium*: cellular and molecular biology, 2nd edn. ASM Press, Washington DC, pp 2067–2117

Colour Figures

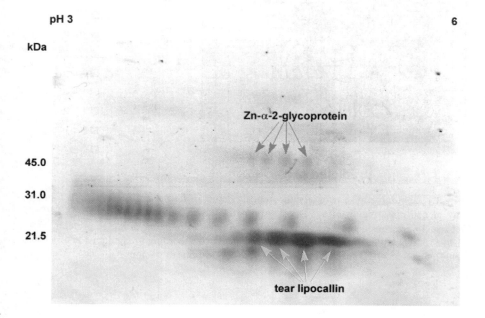

Fig. 2.4. Micropreparative mini 2-D PAGE gel of Human reflex tears. Three μl of tears were diluted to 150 μl with 8 M urea, 4% CHAPS, 2 mM TBP, 40 mM Tris. The 6 cm long, pH 3–6 IPG gel was rehydrated with the whole 150 μl and IEF was performed for 1 hour at 300 V, 30 minutes at 1,000 V, 30 minutes at 2,500 V and 2 hours at 5,000 V. The second dimension was a 8 × 10 cm, 4–20% Bio-Rad mini SDS-PAGE gel. The finished 2-D PAGE gel was stained using 0.1% Coomassie Blue G250 in 12.5% trichloroacetic acid. The entire gel run from IPG rehydration to staining was completed in less than 9 hours

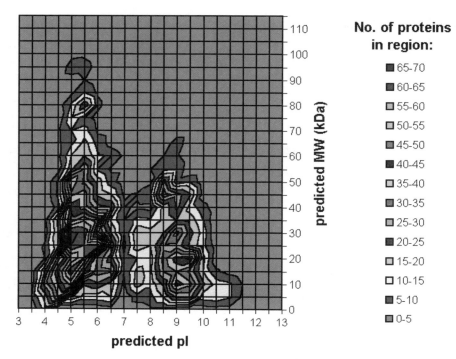

Fig. 3.1. Contour map showing the theoretical number of *Escherichia coli* proteins found in any pI and mass region of a 2-D gel. Approximately 3,500 proteins from SWISS-PROT release 34 are considered here, including *E. coli* plasmid proteins, although proteins larger than 115 kDa are not shown. Protein pI was calculated according to Bjellqvist et al. (1993, 1994). See Sect. 3.3.3 for detail concerning the applications of contour maps

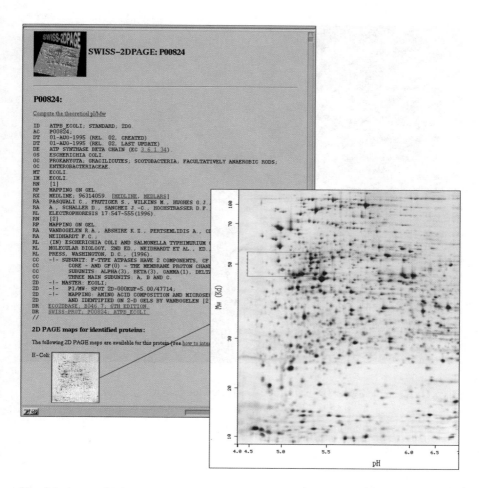

Fig. 5.9. A sample SWISS-2DPAGE entry: the text for ATPB_ECOLI and the master *Escherichia coli* gel to which it is linked

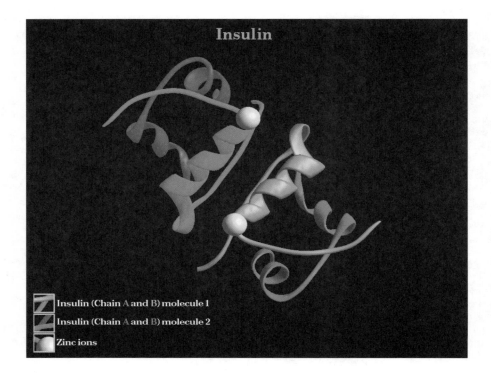

Fig. 5.11. A sample Swiss-3Dimage; the 3-D structure of insulin

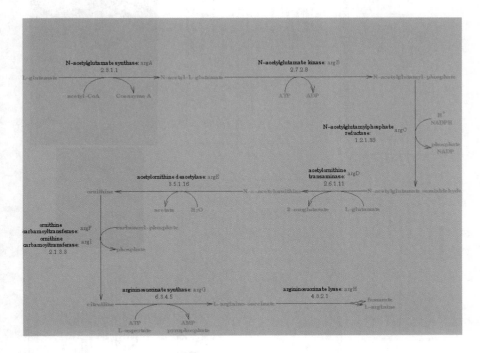

Fig. 5.16. A sample metabolic pathway in EcoCyc

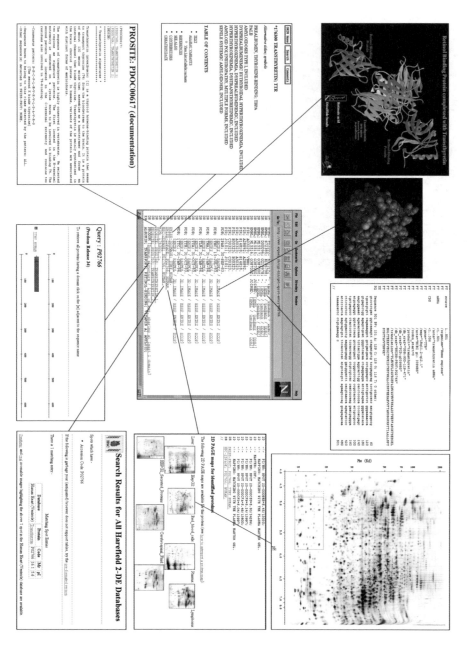

Fig. 6.11. The cross-reference section of the TTHY_HUMAN SWISS-PROT entry, with active links to EMBL/GenBank/DDBJ, PDB, SWISS-3DIMAGE, SWISS-2DPAGE, HSC-2DPAGE, MIM, PROSITE and ProDom

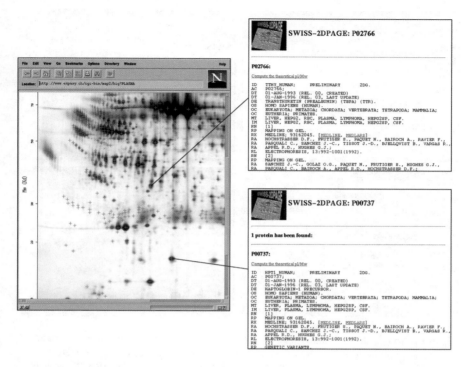

Fig. 6.12. The clickable human plasma protein map from SWISS-2DPAGE as well as examples of two entries that have been retrieved by clicking on the corresponding spot

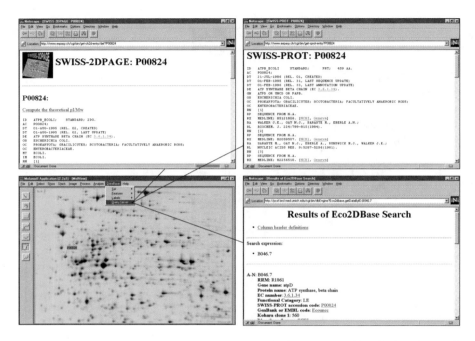

Fig. 6.13. An *E. coli* 2-DE image analysed by the Melanie II 2-D PAGE analysis software. A spot corresponding to ATP synthase beta chain (ATPB_ECOLI) has been selected, and the corresponding entries in SWISS-PROT, SWISS-2DPAGE and ECO2DBASE have been retrieved using the *Query Database* menu item

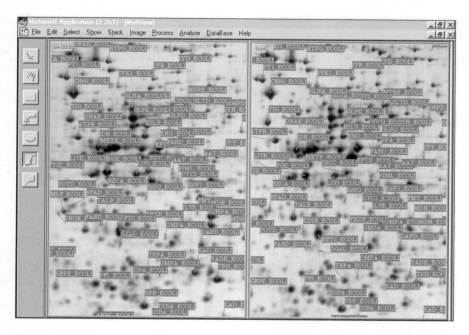

Fig. 6.15. Identified proteins on the *E. coli* 2-DE image of Fig. 6.14 are marked with a label. The gel is then matched to another *E. coli* master, to propagate additional protein identifications

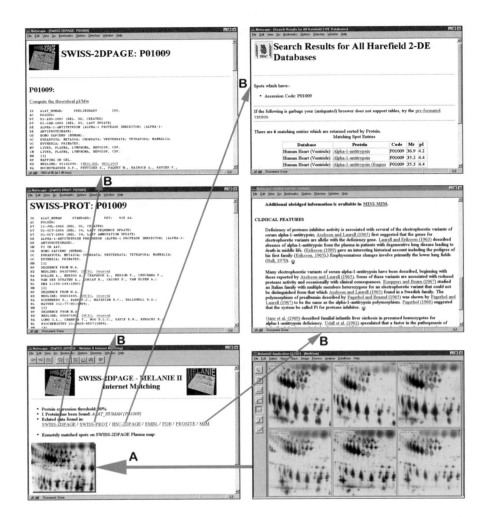

Fig. 6.18. Fictional example of how 2-DE database integration may work in the future: six 2-DE images are analysed by the Melanie II software package (bottom right) that compares them over the Internet to the *Plasma* map in SWISS-2DPAGE. It automatically extracts differentially expressed proteins between two populations (A). Related data is then retrieved from the SWISS-2DPAGE and HSC-2DPAGE 2-DE databases, as well as from the SWISS-PROT protein sequence database and the OMIM on-line Mendelian Inheritance in Man database (B)

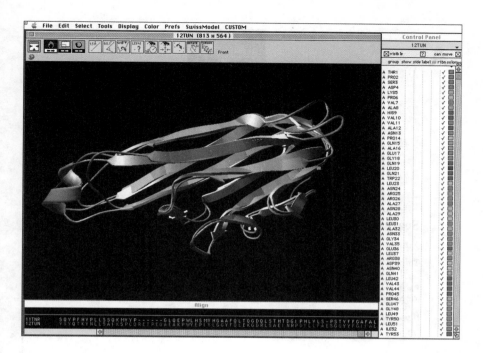

Fig. 7.1. Displaying protein structures with the Swiss-PdbViewer. The 3-D structures of the tumour necrosis factor α and β monomers were loaded into the Swiss-PdbViewer and superposed automatically using the "Magic Fit" option. This screen copy shows the control panel (on the right hand side), from which residues can be selected and display options such as residue colouring, van der Waals surfaces, residue labels and ribbons can be altered. The bottom window holds a structurally-corrected alignment of the two structures. The colour coding of the residues in the bottom sequence reflects their relative mean positional deviation (colours ranging from blue to red) as compared to the top sequence, while the top sequence is depicted grey. This panel can also be used to select residues for which one would like to alter the display options

A

SWISS–MODEL Repository

A database of automatically generated protein models

Search by keyword

This is a searchable index. Enter search keywords: | ODP2 ECOLI

Please enter a keyword. This may be any word or partial word. For example, you may type *dehydrogenase*, or just *dehydro*, or *ODP2_ECOLI* or *odp2*.

If you enter more than one keyword, then only the entries having all keywords will be listed.

Graphical example

If you have problems or comments...

 *Back to the **ExPASy** molecular biology server home page*

B   Search the SWISS–MODEL Repository for: ODP2 ECOLI

Please choose one of the following entries:

ODP2 ECOLI
      DIHYDROLIPOAMIDE ACETYLTRANSFERASE COMPONENT (E2) OF PYRUVATE
      DEHYDROGENASE COMPLEX (EC 2.3.1.12). glycolysis; transferase; acyltransferase;
      repeat; lipoyl.

Fig. 7.2a. The SWISS-MODEL Repository World-Wide Web interface, and the steps required to download and view a model structure. The first step (panel A) involves the downloading of the top page of the SWISS-MODEL Repository from the ExPASy molecular biology server (http://www.expasy.ch/swissmod/swmr-top.html). Keywords such as *ECOLI* or *kinase* can then be used to search the model index. Alternatively, a SWISS-PROT accession or identification code (e.g. ODP2_ECOLI) can be used for direct access to any model in the repository. The results of the query are displayed in a new page which provides links to the actual coordinate entries (panel B)

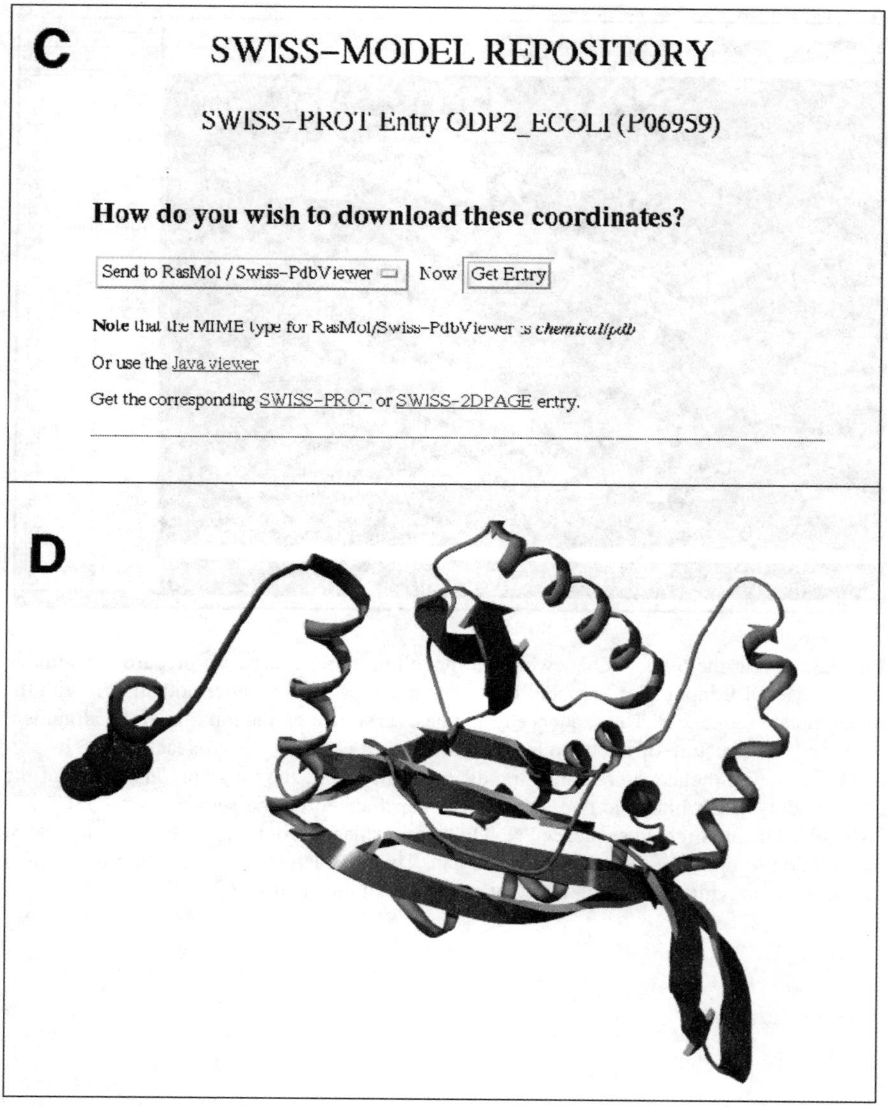

C

SWISS–MODEL REPOSITORY

SWISS–PROT Entry ODP2_ECOLI (P06959)

How do you wish to download these coordinates?

Send to RasMol / Swiss–PdbViewer ⬚ | Now | Get Entry |

Note that the MIME type for RasMol/Swiss–PdbViewer is *chemical/pdb*

Or use the Java viewer

Get the corresponding SWISS–PROT or SWISS–2DPAGE entry.

D

Fig. 7.2b. The SWISS-MODEL Repository World-Wide Web interface, and the steps required to download and view a model structure. The model coordinates (panel C) can be saved to disk or directly loaded into viewing software such as the Swiss-PdbViewer (panel D) or RasMol (Sayle and Milner-White 1995), or displayed and manipulated using a viewer written in Java by Dirk Walther (Table 7.1)

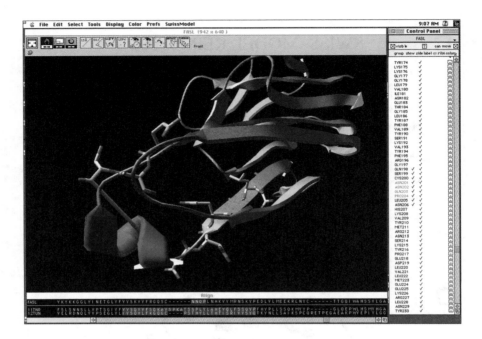

Fig. 7.3. Using the Swiss-PdbViewer as a modelling tool. In order to prepare the suitable SWISS-MODEL input files to model FasL, one must load its sequence and align it with the two template sequences. The sequence of the FasL is shown on the top line of the alignment window. The structure of the tumour necrosis factor β is colour-coded according to its secondary structure (helices in red and strands in yellow). Note the long "pseudo bond" of the FasL model, which links the two sides of the gap flanked by the sequences "GQSC" and "NNQP". The highlighted sequence "NNQP" was displaced from the left to the right side of the deletion to optimise the position of the gap. This was achieved by selecting the desired residues and moving them using the right arrow key of the keyboard

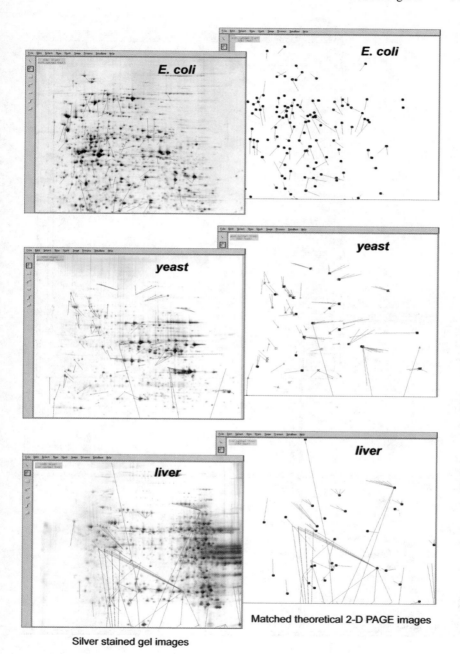

Silver stained gel images

Matched theoretical 2-D PAGE images

Fig. 8.4. Post-translational vector map. Computer matching between theoretical 2-D PAGE maps and the corresponding silver stained gel images from experimental 2-D gels. The vectors highlight the differential migration behaviour of the proteins that are probably due to their post-translational modifications. Modified from Wilkins et al. (1996b)

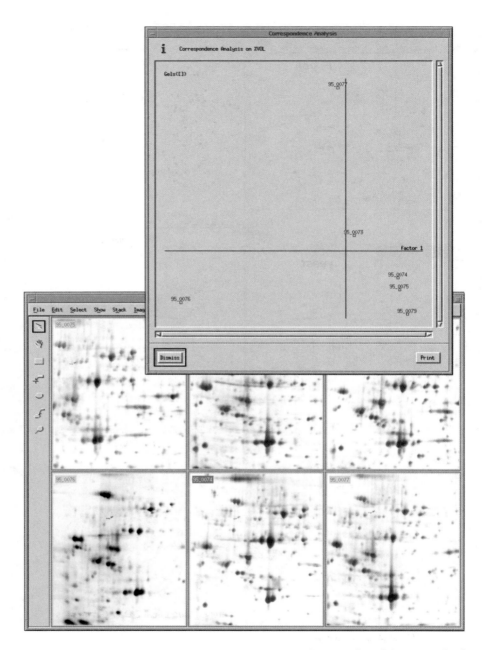

Fig 8.5. Principle component analysis. Projection on the first two factorial spaces of gel images from glial (95_0073, 95_0076 and 93_0077) and melanoma cells (95_0074, 95_0075 and 93_0079). Note that the three melanoma gels are grouped together

6 Interfacing and Integrating Databases

Ron D. Appel

6.1 Introduction

Proteome databases have existed for many years, as have genome databases. As we have seen in Chap. 5, there are currently hundreds of general or specialised proteome databases that are being used by researchers throughout the scientific world. They can be queried, and information can be retrieved that is relevant to the scientist's own experiments. Databases are usually maintained and updated by their authors, and are distributed to the scientific community in various ways.

While each database can be valuable to the individual scientist, two conditions augment the usefulness of proteome databases: when the databases are at all times up-to-date, and when they are linked to each other and may be accessed in an integrated manner. The set of existing proteome databases, when considered jointly, may be regarded as the repository of most of the proteomic knowledge currently available. Present computer and network technology, and especially the Internet and the World-Wide Web (WWW) lets us link together all this huge amount of biological data. It allows us to navigate from one piece of information in one database to entries in other databases, and to connect data about a studied object (being a bacteria, cell line, a patient in a hospital, or just a specific protein or gene) to related data in the various proteome databases, or to its protein map (see Chap. 2). Also, analytical and software tools may be called remotely, such as proteome analysis robots, data analysis packages, tools to quantitatively investigate protein expression levels in given situations, or computer assisted database building systems, in order to push the experimentation a step further.

One may thus state that the proteome analysis techniques detailed in the first four chapters of this book, combined with proteome databases, computer networks and other software packages, form together what we can call a vast *Cyber-Encyclopaedia of the Proteome*, which is at all times readily accessible and hopefully up-to-date. This chapter provides an overview of past and present technologies that can access proteome databases, and shows how networks such as the Internet enable scientists to access and share most of the knowledge available in this Encyclopaedia.

6.2 Past data integration techniques (3300 B.C. − 1993 A.D.)

A database may be regarded as a collection of pieces of information, all related to a given subject or domain. The database is used either as a storage device to retain the information for retrieval at a later point in time, or as a means of disseminating the information by distributing it to other people. From this point of view, databases have existed long before computers. They have been used ever since humans have considered that sharing knowledge with other human beings represented an important and valuable activity. The oldest means of disseminating knowledge is probably speech; this has long been the main way to transmit tradition from one generation to another. Then came writing as a more stable and durable way of collecting and preserving information. It was first employed for this purpose as a means of controlling trade accounting. Around 3300 B.C. tables of numbers were used to record accounting related to various cereal types and products (Nissen et al. 1990). Fig. 6.1 shows part of a clay table written around 3200 B.C. that contains a detailed account of barley distribution. While speech and writing are not (to our standards at the end of the 20th century) the most powerful methods to quickly propagate rapidly changing information, they are to date the only data transmission techniques that have proven efficient over long periods of time. Oral tradition is still the method of choice to convey knowledge from parents to children. Similarly, the Bible is the most notable example of information that has been transmitted nearly unmodified over several millennia to a massive number of people. Its oldest parts written about 3200 years ago, nowadays nearly two billion human beings

Fig. 6.1. Part of a clay table written around 3200 B.C. It contains a detailed account of barley distribution with the totals on the reverse side. Size: 60 × 55 × 18mm. Courtesy Fondation Bodmer, Geneva

have (or are supposed to have) retrieved information from it. And the Dead Sea Scrolls[1], written between the third Century B.C. and 68 A.D., have shown that the currently distributed version of the Bible has practically remained unchanged over the millennia (Dimant and Rappaport 1992).

Biological data were also first disseminated in printed form. In the early days of molecular biology, when databases were growing slowly and when scientists were not yet routinely using computers to communicate, data were collected into books that were called encyclopaedias or atlases (Dayhoff 1978). These could act as data repositories while the rate of change was low, but the pace at which databases now grow makes paper copies impractical as a means to distribute genomic and proteomic databases. Also, it is obvious that printed data lack the flexibility currently required to perform tasks such as data analysis and comparison, or data circulation.

Modern technologies exist that are certainly more appropriate to disseminate data in the current telecommunication era, such as tapes, large hard disks or CD-ROMs. But we currently lack the necessary experience to evaluate for how long it will be possible to retrieve data stored on these media, and to judge whether electronic data will be able to cross millennia as did printed data. Nevertheless, the speed at which new scientific data (and in particular proteomic data) are produced, and the large number of people who ask to obtain this data within a time frame that gets smaller and smaller, makes the Internet the only currently available technology that is appropriate for disseminating proteome databases. This section gives a brief overview of the techniques that were used to integrate proteome databases over the Internet before 1993, when the World-Wide Web became a general purpose data communication tool.

6.2.1 Data exchange on the Internet before WWW

In 1969 the Defense Advanced Research Projects Agency (DARPA), a research agency within the American Department of Defense, inaugurated ARPANET, an experimental nation-wide network. ARPANET rapidly evolved to an international network of computers. In 1977 the TCP/IP protocol was adopted as the main communication protocol to connect to ARPANET. The military segment separated in 1983 and the network then became known as the Internet, that currently links computers all over the world. Until the beginning of the World-Wide Web's (WWW) popularity in 1993 (see Sect. 6.3.2) the Internet was mostly used for world-wide data exchange between people in academic institutions.

In particular, researchers used the various Internet facilities to access and exchange molecular biology databases (Harper 1994). Until the late 1980s, there were mainly three ways of accessing databases over the net: i) electronic mail serv-

---

[1] The Dead Sea Scrolls were discovered in the Judean Desert in 1947 by a young Bedouin shepherd.

ers (Henikoff 1993), ii) FTP and iii) TELNET servers (Comer 1988). Electronic mail servers allow people to retrieve individual entries from databases by sending a detailed query by electronic mail to the mail server's Internet address. The query is then processed by the server, and the result is sent back to the sender's electronic mailbox. This method lets people access individual pieces of data without the need to download the whole database from the net. However, formulating the query can be cumbersome and subject to frequent errors, because it has to comply with a precise syntax. Also, it is a relatively slow database accession technique, as it may take from minutes to hours to obtain the result. On FTP (from File Transfer Protocol) servers, whole databases can be downloaded, and can then be searched on a local machine. While it might be convenient for a large institution to have a local copy of each database that people can access locally, thus limiting the network load, the whole data file has to be downloaded again each time a database has been updated. Finally, TELNET allows a user to remotely log onto a computer and access its facilities, as if he or she was working locally. This protocol assumes of course that the researcher owns a user identification and password on the remote computer. TELNET is useful for occasional database queries, but it usually does not provide any graphical user interface, it requires extensive management of user identifications, and often overloads the remote computer's processing power.

At the turn of the 1990s, two new protocols increased the selection of data accession schemes on the Internet: GOPHER and WAIS (Comer 1988). Through GOPHER, data on a server can be organised in a tree-like form, and one can navigate the tree until the pertinent piece of information has been found. Lacking a graphical user interface, this protocol has rapidly been replaced by the WWW. Using WAIS (from Wide Area Information Server) a full inverted index can be constructed from a large number of text files, that may then be searched over the Internet, and individual files retrieved. Also partially replaced by the WWW paradigm, WAIS is still being used in given situations, such as full text searches on large databases.

6.2.2 Hard media data distribution

As many scientists do not have access to a high speed connection to the Internet, major proteome databases are still being distributed on hard storage media. While the floppy disk used to be employed for this purpose, it is not any more the medium of choice to store databases, because proteome databases tend to be too large.[2] Some organisations, such as the European Bioinformatics Institute (EBI) at Hinx-

---

[2] For example, the SWISS-PROT protein sequence database, release 34 of January 1997 (Bairoch and Apweiler 1997) takes up 43 MB of compressed disk space, and would require no less than 30 floppy disks.

ton, U.K., now distribute the EMBL nucleotide sequence database not only on the Internet, but also on CD-ROMs.[3]

6.2.3 Local database integration

As mentioned in Sect. 6.2.1, the FTP protocol is used to download full copies of databases in order to access their content on local machines. Specialised software programs have been developed that enable scientists to query either individual or locally integrated databases, i.e. local copies of several databases accessible through a unique, usually custom-made user interface. Among the large number of public domain or commercial software packages that are available, the GCG Suite (Devereux et al. 1984) is one of the well known sequence analysis packages that accesses local versions of several proteome databases.

Another type of computer software that integrates databases locally involves programs that provide a uniform user interface to a number of databases stored locally on one server, and also make this interface accessible on the Internet. The Sequence Retrieval System (SRS) of Etzold and Argos (1993) provides an integrated World-Wide Web access[4] (see Sect. 6.3) to more than 40 proteome and genome databases that are all stored and indexed locally on the EBI server. Through a single query, SRS returns data from any of the indexed databases.[5] A second important software program that integrates databases locally is Entrez (Schuler et al. 1996) developed at the National Center for Biotechnology Information (NCBI) of the National Library of Medicine (NLM), at the National Institutes of Health (NIH), Bethesda, Maryland. Integrating nine databases — including GenBank (Benson et al. 1996), dbEST (Rodriguez-Tome 1997), SWISS-PROT (Bairoch and Apweiler 1997), GSDB (Harger et al. 1997) and the NLM's MEDLINE bibliographical reference database — the main goal of Entrez is to search the molecular biology subset of MEDLINE, and to provide links to related data in the other eight indexed databases. Noteworthy is a feature called *document neighbours* that produces pointers to other bibliographical references that are similar to the query, according to a text similarity measure developed by Wilbur and Yang (1996). Entrez is also accessible on the Internet.[6]

---

[3] The EBI distribution of the EMBL release 50 of April 1997 was stored on 6 CD-ROMs.

[4] http://srs.ebi.ac.uk:5000/

[5] It is worth mentioning that this scheme, where one WWW server gives access to several locally kept databases, is different from the general WWW paradigm that will be discussed in Sect. 6.3, according to which a unique user interface lets scientists query databases that are distributed over a large number of servers world-wide. Keeping local copies of databases that are maintained at other sites raises the problem of guaranteeing that the databases are at all times up-to-date.

[6] http://www3.ncbi.nlm.nih.gov/Entrez/

6.2.4 Discussion of standard techniques

In addition to the advantages and weaknesses of conventional database distribution methods that have already been exposed in the text, such as the lack of flexibility of hard storage media, a few other aspects have to be addressed. Apart from TELNET, all techniques described in the previous sections have in common the fact that they function on full copies of the databases. This has the consequence that each time one wants to update one of the local databases to a newer release, the full database has to be downloaded. Thus, besides the database taking up a considerable amount of disk space, the process of updating to the new release can be time consuming.

Furthermore, a locally integrated multi-database system, being part of a software package dedicated to local use only, or of an Internet service such as SRS or Entrez, is in most cases maintained by a system or database administrator. The user, when locally or remotely querying the databases, has not always the assurance that he or she is working on the most recent version of the database. Cases have been seen where a Web server was returning results from a two year old release of the SWISS-PROT protein sequence database. Finally, all major proteome databases, in addition to being subject to regular updates (usually every one or two months) in the form of new releases, also undergo partial, weekly or even daily updates, that are not incorporated in the main releases. These updates will therefore not be included in local copies of the databases.

6.3 Current integration: a *Cyber-Encyclopaedia of the Proteome*

As stated in this chapter's introduction, the hundreds of proteome databases that are being maintained around the world jointly hold most of the currently available proteomic knowledge. Being able to extract up-to-date data from these databases in an integrated and efficient way represents a key asset to each scientist working on a subject related to the proteome. This section shows how current technology allows us to interface and integrate most proteome databases in such a way that the information pertinent to a given topic may be readily consulted, and that data are guaranteed to be completely up-to-date. Additionally, this may be achieved without the need to download full copies of databases, with only the subject of interest being retrieved at any time over the net. It thus forms a vast world-wide *Cyber-Encyclopaedia of the Proteome*.

6.3.1 Cross-references

The most elementary link from one database to another one is the cross-reference. Using this principle, references may be given in database entries to related entries in other databases. This scheme is being used extensively in the SWISS-PROT protein sequence database (Sect. 5.2.1; Bairoch and Apweiler 1997) in the DR (Database cross-Reference) lines. The format of a DR line, except for cross-references to the EMBL/Genbank/DDBJ nucleotide sequence database, is the following:

```
DR    DATABASE_ID; PRIMARY_ID; SECONDARY_ID.
```

DATABASE_ID contains the database identifier, that is the name of the database that holds the related entry. PRIMARY_ID is the entry's primary key (often called the accession number), while SECONDARY_ID complements the information given by the first identifier. The format of a cross-reference to the EMBL/Genbank/DDBJ database is:

```
DR    EMBL; ACCESSION_NUMBER; PID; STATUS_ID.
```

ACCESSION_NUMBER is the EMBL entry's primary key, PID is the Protein IDentification number, and STATUS_ID provides information about the relationship between the sequence in the SWISS-PROT entry and the coding sequence in the corresponding EMBL entry. Fig. 6.2 shows the cross-reference portion of the human alpha-1 antitrypsin entry in the SWISS-PROT database. It provides links to related entries in the EMBL nucleotide sequence database (Rodriguez-Tome et al. 1996), the PIR Protein Information Resource (George et al. 1996), the PDB X-ray crystallography Protein Data Bank (Abola et al. 1996), the SWISS-3DIMAGE database of annotated 3-D images (Peitsch et al. 1995), the SWISS-2DPAGE database of two-dimensional gel electrophoresis (2-DE) images (Appel et al. 1996b), the HSC-2DPAGE 2-DE database (Corbett et al. 1994), the MIM Mendelian Inheritance in Man database (Pearson et al. 1994) and the PROSITE dictionary of sites and patterns (Bairoch et al. 1997). This scheme lets users search through cross-referenced databases in order to retrieve part or all of the related information. While cross-references are now widely utilised in most genome and proteome databases, they present a number of significant limitations. Firstly, cross-references are only unidirectional. If a database entry has a cross-reference to an entry in another database, there is no assurance that the latter shall contain a pointer to the former entry. Secondly, accession to all entries that cross-references point to can be laborious. For example, the serine proteinase inhibitors entry in the PROSITE sites and patterns database contains cross-references to the SWISS-PROT entries of 111 proteins belonging to this group of structurally related proteins, or presenting their signature.[7] Querying the database simultaneously for all information contained in the 111 entries is cumbersome or impossible (Fig. 6.3). Third, performing complex

---

[7] 85 true positives (proteins belonging to this protein family, and whose sequences contain the serpins signature), 14 false positives (proteins not belonging to the serpins, but holding the signature), 6 false negatives, and 6 partial (sequences belonging to the protein family, but do not completely match the pattern, because they are partial sequences).

```
DR    EMBL; V00496; G825632; -.
DR    EMBL; K01396; G177829; -.
DR    EMBL; X01683; G28966; -.
DR    EMBL; J00064; G177822; -.
DR    EMBL; J00066; G177823; -.
DR    EMBL; J00065; G177823; JOINED.
DR    EMBL; J00067; G177824; -.
DR    EMBL; J02619; G177836; -.
DR    EMBL; X02920; G24438; -.
DR    EMBL; K02212; G177831; -.
DR    PIR; A21853; ITHU.
DR    PDB; 7API; 15-OCT-90.
DR    PDB; 8API; 15-OCT-90.
DR    PDB; 9API; 15-OCT-90.
DR    SWISS-3DIMAGE; A1AT_HUMAN.
DR    SWISS-2DPAGE; P01009; HUMAN.
DR    HSC-2DPAGE; P01009; HUMAN.
DR    MIM; 107400; -.
DR    PROSITE; PS00284; SERPIN.
```

Fig. 6.2. Cross-references in the human alpha-1 antitrypsin entry of the SWISS-PROT protein sequence database

queries through the cross-references is not possible. For example, to obtain only those entries from the above mentioned 111 cross-references that contain a given keyword, one would have to retrieve all entries and search them individually for the keyword. Fourth, not all genome and proteome databases have been comprehensively cross-referenced. For instance, each SWISS-2DPAGE entry possesses cross-references to other 2-DE databases and to SWISS-PROT, but it does do not include direct references to other databases cross-referenced in SWISS-PROT. Despite these restrictions, the cross-reference still constitutes the main component of database integration, when it is complemented with efficient cross-database navigational tools. These tools are supplied by the World-Wide Web paradigm.

6.3.2 The World-Wide Web

The World-Wide Web (WWW) was invented in 1990 by Tim Berners-Lee at CERN, Geneva, Switzerland, to facilitate data exchange between physics researchers (Berners-Lee et al. 1992). It is based on the hypertext concept according to which a text may be displayed, and given elements (a word, sentence or part of a sentence) highlighted. Each highlighted element, called an *anchor*, is a pointer to another document with similarly anchored components. A set of such texts may form a large network of interlinked documents and is called a *hypertext*. On computers that are equipped with a graphical display and a mouse device, an anchored element is often highlighted in colour, and by clicking on it with the mouse, the

```
ID   SERPIN; PATTERN.
AC   PS00284;
DT   APR-1990 (CREATED); JUN-1994 (DATA UPDATE); NOV-1995 (INFO UPDATE).
DE   Serpins signature.
PA   [LIVMFY]-x-[LIVMFYAC]-[DNQ]-[RKHQS]-[PST]-F-[LIVMFY]-[LIVMFYC]-x-
PA   [LIVMFAH].
NR   /RELEASE=32,49340;
NR   /TOTAL=99(99); /POSITIVE=85(85); /UNKNOWN=0(0); /FALSE_POS=14(14);
NR   /FALSE_NEG=6; /PARTIAL=6;
CC   /TAXO-RANGE=??E?V; /MAX-REPEAT=1;
DR   P01011, AACT_HUMAN, T; P22323, COTR_CAVPO, T; P07759, COTR_MOUSE, T;
DR   P22324, A1AF_CAVPO, T; P22325, A1AS_CAVPO, T; P01009, A1AT_HUMAN, T;
DR   P20848, A1AU_HUMAN, T; P01010, A1AT_PAPAN, T; P38028, A1A1_HORSE, T;
DR   P38029, A1A2_HORSE, T; P38031, A1A4_HORSE, T; P07758, A1A1_MOUSE, T;
DR   P22599, A1A2_MOUSE, T; P26595, A1AT_MUSCR, T; P23035, A1AF_RABIT, T;
DR   P17475, A1AT_RAT  , T; P34955, A1AT_BOVIN, T; P12725, A1AT_SHEEP, T;
DR   P32759, A1AT_CYPCA, T; P22922, A1AT_BOMMO, T; Q03383, ACH1_BOMMO, T;
DR   P80034, ACH2_BOMMO, T; P28800, A2AP_BOVIN, T; P08697, A2AP_HUMAN, T;
DR   P05155, IC1_HUMAN , T; P41361, ANT3_BOVIN, T; P01008, ANT3_HUMAN, T;
DR   P32261, ANT3_MOUSE, T; P32262, ANT3_SHEEP, T; P01019, ANGT_HUMAN, T;
DR   P07093, GDN_HUMAN , T; P07092, GDN_RAT   , T; P13909, PAI1_BOVIN, T;
DR   P05121, PAI1_HUMAN, T; P22777, PAI1_MOUSE, T; P20961, PAI1_RAT  , T;
DR   P05120, PAI2_HUMAN, T; P12388, PAI2_MOUSE, T; P29524, PAI2_RAT  , T;
DR   P05154, IPSP_HUMAN, T; P05543, THBG_HUMAN, T; P35577, THBG_RAT  , T;
DR   P08185, CBG_HUMAN , T; Q06770, CBG_MOUSE , T; P23775, CBG_RABIT , T;
DR   P31211, CBG_RAT   , T; P01012, OVAL_CHICK, T; P19104, OVAL_COTJA, T;
DR   P01013, OVAX_CHICK, T; P01014, OVAY_CHICK, T; P06293, PRTZ_HORVU, T;
DR   P29621, KBP_MOUSE , T; P05545, KBP_RAT   , T; P05544, SPI1_RAT  , T;
DR   P09005, SI21_RAT  , T; P09006, SPI3_RAT  , T; P05546, HEP2_HUMAN, T;
DR   P05619, ILEU_HORSE, T; P30740, ILEU_HUMAN, T; P80229, ILEU_PIG  , T;
DR   P29508, SCCA_HUMAN, T; P35237, PTI_HUMAN , T; P12393, SPI1_MYXVL, T;
DR   P20531, SPI1_VACCC, T; P15058, SPI1_VACCV, T; P33829, SPI1_VARV , T;
DR   P42927, SPI1_COWPX, T; P42928, SPI1_RABPU, T; P20842, SPIB_VACCC, T;
DR   P15059, SPI2_VACCV, T; P33830, SPI2_VARV , T; P42926, SPI2_RABPU, T;
DR   P07385, CRMA_COWPX, T; P14754, SERA_MANSE, T; P46201, UTMP_BOVIN, T;
DR   P21814, UTMP_SHEEP, T; P16708, UFBP_PIG  , T; P46202, UAB2_PIG  , T;
DR   Q00387, EP45_XENLA, T; P13731, HS47_CHICK, T; P29043, HS47_HUMAN, T;
DR   P19324, HS47_MOUSE, T; P29457, HS47_RAT  , T; P36952, MASP_HUMAN, T;
DR   P36955, PEDF_HUMAN, T;
DR   Q03352, ANT3_CHICK, P; P01017, ANGT_BOVIN, P; P01018, ANGT_CHICK, P;
DR   P01016, ANGT_HORSE, P; P20757, ANGT_SHEEP, P; P29622, KAIN_HUMAN, P;
DR   Q03044, A1AT_DIDMA, N; P11859, ANGT_MOUSE, N; P01015, ANGT_RAT  , N;
DR   P20532, SPI3_VACCC, N; P18384, SPI3_VACCV, N; P33831, SPI3_VARV , N;
DR   P43634, CHA4_YEAST, F; P37126, CGLH_XANMA, F; Q01886, HTS1_COCCA, F;
DR   P18654, KS62_MOUSE, F; P44961, MENC_HAEIN, F; P15577, NU2M_PARTE, F;
DR   P10329, NU6M_CHLRE, F; P13899, POLG_TMEVD, F; P15324, STSY_RAUSE, F;
DR   P38369, TP50_TREPA, F; Q01003, UL79_HSVSA, F; P26541, VL3_HPV5B , F;
DR   P12221, YCF0_MARPO, F; Q04855, YNTC_AZOCA, F;
3D   2ACH; 7API; 8API; 9API; 1PAI; 2PAI; 1OVA; 1HLE;
DO   PDOC00256;
//
```

Fig. 6.3. The Serpins signature entry in the PROSITE database

pointed-to document can automatically be retrieved and displayed (Fig. 6.4). A simple hypertext language, HTML (from HyperText Mark-up Language) has been designed for the WWW. Using HTML's tagged language, complex hypertext documents may be described in text (ASCII) format. This makes HTML documents independent of hardware and operating system. Furthermore, the simplicity of the HTML syntax[8] makes it easy to design software for any computer environment

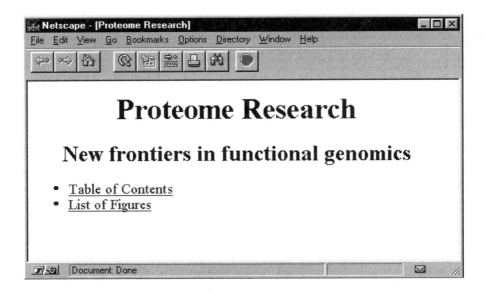

Fig. 6.4. A simple hypertext document; anchors, which represent hypertext links, are underlined

that is able to display hypertext documents (Fig. 6.5). The WWW also defined a new Internet protocol, HTTP (from HyperText Transfer Protocol) that optimises transmission of HTML documents. The portability of HTML, combined with the efficiency of the HTTP protocol to transmit documents over the Internet, transformed the basic hypertext concept into a global, world-wide network of hypertext documents, as each anchored element in a HTML document may be the pointer to another document, not only on the user's local computer, but on any computer linked to the Internet. A piece of software capable of displaying HTML documents and of retrieving other HTML documents over the Internet through simple mouse clicks is called a *browser*.

In 1993 Marc Andreessen at the National Center for Supercomputing Applications (NCSA), Champaign, Illinois, extended the WWW by developing Mosaic, the WWW's first graphical browser. With this addition, the WWW became a world-wide hypermedia network, able to display HTML documents containing not only text, but also images, animated image sequences, as well as sound. Currently, many browsers are available for most computer hardware and operating environments. These include Mosaic[9], Netscape Navigator[10] and Internet Explorer[11].

---

[8] http://www.w3.org/hypertext/WWW/MarkUp/MarkUp.html

[9] http://www.ncsa.uiuc.edu/SDG/Software/Mosaic/NCSAMosaicHome.html

```
<html>
<head>
<TITLE>Proteome Research</TITLE>
</head>
<body>
<center><H1>Proteome Research</H1></center>
<center><H2>New frontiers in functional genomics</H2>
</center>
<UL>
<LI><A HREF=toc.html>Table of Contents</A>
<LI><A HREF=figs.html>List of Figures</A>
</UL>
</body>
</html>
```

Fig. 6.5. The hypertext document of Fig. 6.4 represented in HTML

6.3.3 Integrated databases on the World-Wide Web

Shortly after the Mosaic browser was made publicly available by the NCSA, it became clear that the World-Wide Web (WWW) paradigm would greatly enhance the power of cross-references in databases by providing active integration of databases over the Internet, thus eliminating the need to download and maintain local copies of databases. Furthermore, it would give the scientist the ability to easily navigate across database entries through active hypertext cross-references, with the guarantee of each retrieved piece of information being at all times up-to-date. The first molecular biology World-Wide Web server, called ExPASy (from Expert Protein Analysis System) was set up at Geneva University Hospital and University of Geneva in August 1993. It allowed queries to four of the databases maintained in Geneva — SWISS-PROT, SWISS-2DPAGE, SWISS-3DIMAGE and PROSITE (Appel et al. 1993, 1994).[12] During the following months most major genome and proteome databases were made accessible on Web servers throughout the world. ExPASy currently offers entry points to three additional local databases — the CD40Lbase (the European CD40L Defect Database), ENZYME (enzyme nomenclature database) and SeqAnalRef (sequence analysis bibliographic reference database) — with hypertext links to more than 20 databases on other Web servers. Access is also provided to on-line sequence analysis tools, software tutorials, an on-line version of Boehringer Mannheim's map of biochemical pathways, proteome-related documents and other pieces of information and services (Hochstrasser et al. 1995; Wilkins et al. 1997a, b).

---

[10] http://home.netscape.com/

[11] http://www.microsoft.com/ie/default.asp

[12] http://www.expasy.ch/

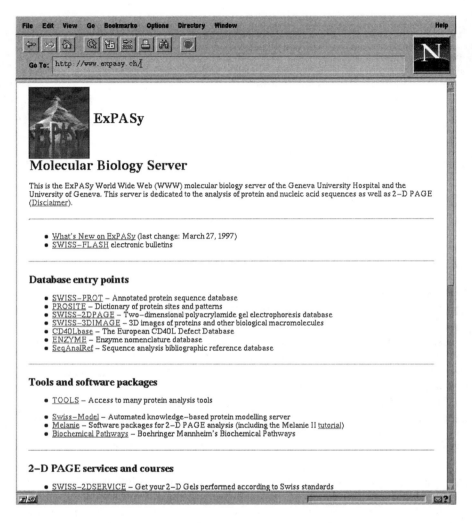

Fig. 6.6. The ExPASy home page on the WWW, as seen with a browser

The integration of databases on the WWW is best illustrated by an example. Fig. 6.6 shows the home page of the ExPASy molecular biology Web server, that gives access to the various local databases and software tools. A mouse click on the underlined reference to SWISS-PROT requests the SWISS-PROT top page (Fig. 6.7) from which one of the database query methods is chosen. Selecting "by description or identification" displays a form requiring one or more keywords to be entered. Typing "transthyretin" activates a search of SWISS-PROT, which returns a list of all 14 currently available entries matching the search criterion (Fig. 6.8).

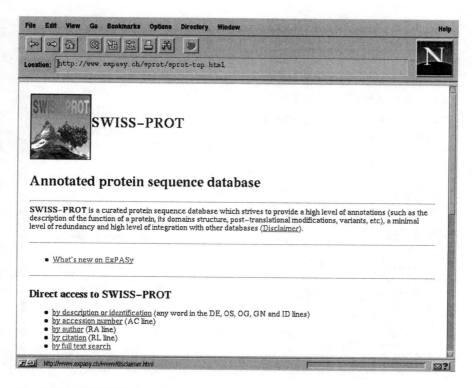

Fig. 6.7. The SWISS-PROT top page

Clicking on TTHY_HUMAN retrieves the SWISS-PROT entry for human trans-thyretin. Fig. 6.9 shows an extract of this entry with bibliographical references and annotations. Fig. 6.10 displays the amino acid sequence and active links to various on-line tools such as Swiss-Model that can produce a three-dimensional model of the protein (Chap. 7 and Peitsch 1996), BLAST sequence alignment, ProtParam that computes physico-chemical parameters of the protein sequence (molecular weight, theoretical pI, amino acid composition, extinction coefficients, estimated half-life, instability and aliphatic index), ProtScale that computes and represents the profile produced by any amino acid scale on the protein, or PeptideMass that cuts the protein sequence with a chosen enzyme and computes the masses of the generated peptides (Wilkins et al. 1997a). Fig. 6.11 (see p. 138) shows the cross-reference section of the TTHY_HUMAN entry, with active links to several other databases, including EMBL/GenBank/DDBJ, SWISS-3DIMAGE, SWISS-2DPAGE, HSC-2DPAGE, MEDLINE, MIM, PROSITE and ProDom (Sonnhammer and Kahn 1994).

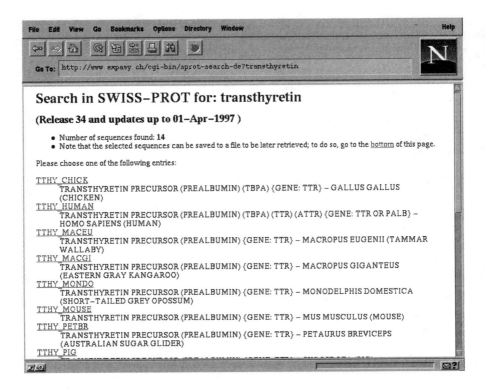

Fig. 6.8. The first 8 of 14 currently available transthyretin entries in SWISS-PROT

6.3.4 Federated proteome databases

The active hypertext link, that lets users navigate the Internet from one database entry to any related entry in another database, forms the basic building block of database integration. *Federated database* is a concept that generalises the use of the active hypertext link to ensure that a database may be queried over the Internet, and that each of its entries is linked to relevant data in other federated databases (Appel et al. 1996a). Federated databases may be developed and maintained independently at various laboratories, using different database technologies. They may be queried separately through the Internet or as stand-alone databases. But they are linked together through the WWW. A federated database has to comply with the following three rules:

Rule 1: Individual entries in the database must be accessible by remote keyword search. This provides an efficient way of querying a database and is in most cases sufficient. Other query methods are possible but not required, such as full text search.

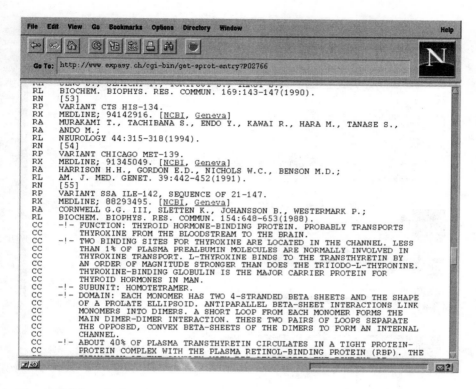

File   Edit   View   Go   Bookmarks   Options   Directory   Window   Help

Go To: http://www.expasy.ch/cgi-bin/get-sprot-entry?P02766

```
RL   BIOCHEM. BIOPHYS. RES. COMMUN. 169:143-147(1990).
RN   [53]
RP   VARIANT CTS HIS-134.
RX   MEDLINE; 94142916. [NCBI, Geneva]
RA   MURAKAMI T., TACHIBANA S., ENDO Y., KAWAI R., HARA M., TANASE S.,
RA   ANDO M.;
RL   NEUROLOGY 44:315-318(1994).
RN   [54]
RP   VARIANT CHICAGO MET-139.
RX   MEDLINE; 91345049. [NCBI, Geneva]
RA   HARRISON H.H., GORDON E.D., NICHOLS W.C., BENSON M.D.;
RL   AM. J. MED. GENET. 39:442-452(1991).
RN   [55]
RP   VARIANT SSA ILE-142, SEQUENCE OF 21-147.
RX   MEDLINE; 88293495. [NCBI, Geneva]
RA   CORNWELL G.G. III, SLETTEN K., JOHANSSON B., WESTERMARK P.;
RL   BIOCHEM. BIOPHYS. RES. COMMUN. 154:648-653(1988).
CC   -!- FUNCTION: THYROID HORMONE-BINDING PROTEIN. PROBABLY TRANSPORTS
CC       THYROXINE FROM THE BLOODSTREAM TO THE BRAIN.
CC   -!- TWO BINDING SITES FOR THYROXINE ARE LOCATED IN THE CHANNEL. LESS
CC       THAN 1% OF PLASMA PREALBUMIN MOLECULES ARE NORMALLY INVOLVED IN
CC       THYROXINE TRANSPORT. L-THYROXINE BINDS TO THE TRANSTHYRETIN BY
CC       AN ORDER OF MAGNITUDE STRONGER THAN DOES THE TRIIODO-L-THYRONINE.
CC       THYROXINE-BINDING GLOBULIN IS THE MAJOR CARRIER PROTEIN FOR
CC       THYROID HORMONES IN MAN.
CC   -!- SUBUNIT: HOMOTETRAMER.
CC   -!- DOMAIN: EACH MONOMER HAS TWO 4-STRANDED BETA SHEETS AND THE SHAPE
CC       OF A PROLATE ELLIPSOID. ANTIPARALLEL BETA-SHEET INTERACTIONS LINK
CC       MONOMERS INTO DIMERS. A SHORT LOOP FROM EACH MONOMER FORMS THE
CC       MAIN DIMER-DIMER INTERACTION. THESE TWO PAIRS OF LOOPS SEPARATE
CC       THE OPPOSED, CONVEX BETA-SHEETS OF THE DIMERS TO FORM AN INTERNAL
CC       CHANNEL.
CC   -!- ABOUT 40% OF PLASMA TRANSTHYRETIN CIRCULATES IN A TIGHT PROTEIN-
CC       PROTEIN COMPLEX WITH THE PLASMA RETINOL-BINDING PROTEIN (RBP). THE
```

Fig. 6.9. Bibliographical references and annotations in the SWISS-PROT entry for human transthyretin

Rule 2: The database must be linked to other databases through active hypertext cross-references. Database entries must have such a cross-reference to at least the main index (see Rule 3).

Rule 3: In addition to individually searchable databases, a main index has to be supplied that provides a means of querying all databases through one unique entry point. Bidirectional cross-references must exist between the main index and the other databases.

The examples given in Sect. 6.3.3 and Fig. 6.11 (see p. 138) demonstrate the power of federated proteome databases. Each of the displayed databases can be interrogated separately over the net. They connect to each other by hypertext links. The SWISS-PROT database serves as the main index: it is searchable by keyword and holds links to each of the databases.

Proteome research begins with the graphical presentation of the protein complement of a cell or tissue through display of 2-DE images. Information in these images has been incorporated into the federated database model. A federated 2-DE

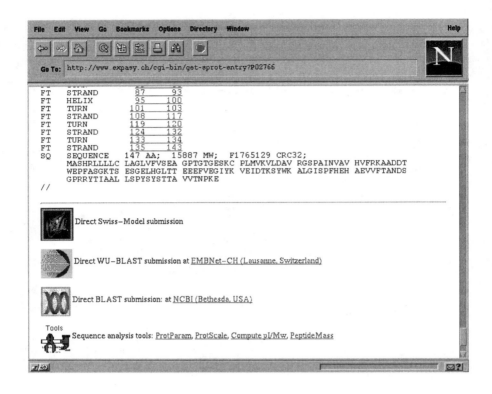

Fig. 6.10. The amino acid sequence of human transthyretin and active links to Swiss-Model, BLAST, ProtParam, ProtScale, Compute pI/Mw and PeptideMass

database must therefore comply with the following additional two rules (Appel et al. 1996a):

Rule 4: Individual protein entries must be accessible through clickable images. That is, 2-DE images must be provided on the WWW server and, as a response to a mouse click on any identified spot on the image, the user must obtain the database entry for the corresponding protein. This method allows a user to easily find identification data for proteins on a 2-DE image. Fig. 6.12 (see p. 139) shows the clickable human plasma protein map from SWISS-2DPAGE (Golaz et al. 1993) as well as examples of two entries that have been retrieved by clicking on the corresponding spots.

Rule 5: 2-DE analysis software designed for use with federated databases must be able to directly access individual entries in any federated 2-DE database. For example, when displaying a 2-DE reference map with a 2-DE image analysis program, the user must be able to select a spot and remotely obtain the corresponding entry from the given database. In Fig. 6.13 (see p. 140) an *E. coli* 2-DE image has been analysed with the Melanie II 2-D PAGE analysis software (Wilkins et al.

1996).[13] A spot has been selected, and the entries of the corresponding protein have been retrieved from three different federated proteome databases.

Tools to implement a federated proteome database exist and are easy to use. Several public domain software packages are available for setting up a WWW server, such as the Apache Web server software from the National Center for Supercomputing Applications (NCSA)[14]. Rule 1 (keyword search) may be implemented using either *fill-out forms* or even simpler using the *isindex* tag. Both features are part of HTML. Rules 2 and 3 may be implemented using the anchor tag in the HTML language. An anchor takes the following form:

```
<A HREF=http://host/dir/prog?ID>Text</A>
```

where *host* is the name of the server on which the remote database is located, *prog* is the program to retrieve the entry, *dir* is the directory in which the latter resides, *ID* is the entry's unique identification and *Text* is the anchor's text which appears in the current entry. For example, the Kyoto University ligand chemical database[15] entry for ATP synthase carries the following active cross-reference to the corresponding ENZYME entry (shown in HTML format):

```
<A HREF=http://www.expasy.ch/htbin/
get-enzyme-entry?3.6.1.34>3.6.1.34</A>
```

Rule 4 may be satisfied by exploiting the *ismap* concept that allows individual regions of an image to be active links to other WWW documents.

Finally, a 2-DE computer analysis software may comply with Rule 5 by remotely controlling a WWW browser and requesting the following document for any given protein:

```
http://host/cgi-bin/get-2d-entry/database?ID
```

where *host* is the name of the server on which the remote database is located, *database* is the selected database on that server, and *ID* is the entry's unique identification. For example, in order to retrieve the ATP synthase entry from the SWISS-2DPAGE database on the ExPASy server, the following request has to be sent by the WWW browser:

```
http://www.expasy.ch/cgi-bin/get-2d-entry/SWISS-2DPAGE?P00824
```

Several 2-DE databases have already been set up on the WWW, and most of them are federated to some degree. An index to existing federated 2-DE databases

---

[13] http://www.expasy.ch/melanie/melanie-top.html

[14] The Apache Web software may be downloaded from the NCSA anonymous FTP server at ftp://ftp.ncsa.uiuc.edu/

[15] http://www.genome.ad.jp/

called WORLD-2DPAGE[16] is available on the ExPASy server and is updated on a regular basis. It also contains the guidelines for building a federated 2-DE database. A general list of genome and proteome databases that are accessible on the WWW is maintained on ExPASy by Amos Bairoch[17] (see Table 5.1).

6.3.5 Discussion of current integration

Many genome and proteome databases developed and maintained by research groups throughout the world are now accessible on the World-Wide Web. The simple mechanism of active hypertext cross-references joins all related databases together, combining them into one large virtual database, a *Cyber-Encyclopaedia of the Proteome*. While much work is still required to resolve issues such as automatic cross-referencing, synchronising database updates, network overload, database distribution, database mirroring and handling of redundant databases, the WWW paradigm currently provides the most efficient way of interfacing and integrating proteome databases.

6.4 Integration of databases and analytical methods

Most proteome databases are currently being established manually, adding data either from results of analytical methods or from the literature. However most sophisticated laboratory devices such as sequencers, mass spectrometers and amino acid analysers are connected to computers, and results are produced in electronic form. Relatively little effort would be required in order to transfer data directly from such equipment to databases. We can thus expect at least partial automation of database construction in the near future (see Sect. 6.5.3). Indeed, the complexity of protein maps produced by 2-DE techniques has already required the development of computer tools to assist investigators in building 2-DE databases. A 2-DE gel on which proteins are to be identified is usually digitised using a laser densitometer, a CCD camera or a phosphor imager (Patton et al. 1997) and read into a 2-DE software analysis program (Wilkins et al. 1996; Appel and Hochstrasser 1997). After optional image processing to enhance the resulting image quality, protein spots are then automatically detected (Fig. 6.14). The identification of proteins on the gel (see Chap. 3) allows spots to be labelled with names or other identifications obtained from protein sequence databases. Such a 2-DE gel image containing identified proteins is often called a *master protein map*. Additional proteins may then

---

[16] http://www.expasy.ch/ch2d/2d-index.html

[17] http://www.expasy.ch/www/amos_www_links.html

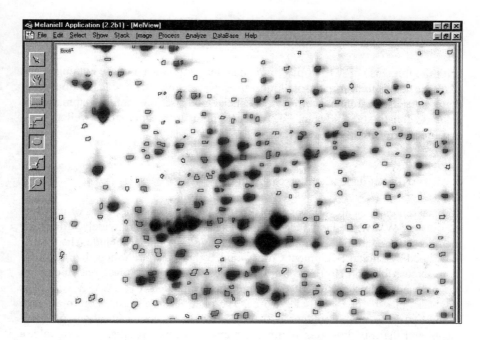

Fig. 6.14. Spot borders detected on an *E. coli* 2-DE image by the Melanie II 2-D PAGE analysis software have been highlighted

be identified by comparison with other existing master maps from the same species, using 2-DE matching algorithms that are part of most 2-DE analysis software (Fig. 6.15, see p. 141). The next step consists of creating new entries in the 2-DE database, or augmenting existing entries with additional data. While this is often performed manually, tools have been created that help construct a 2-DE database. Starting from a master 2-DE gel image containing labels for each identified protein, the *make2ddb* software package produces files and programs that are necessary to set up a WWW accessible 2-DE database (Hoogland et al. 1997; Appel et al. 1997). This public domain software is available on the Internet[18] and helps integrate gel images with 2-DE databases.

---

[18] http://www.expasy.ch/ch2d/make2ddb.html

6.5 The future of proteome database integration

With the immense growth of proteomic data that will happen in the coming years, five aspects of bioinformatics will most probably receive increased attention: intelligent database search and retrieval, enhanced database integration, automation of database construction, advanced data processing over the Internet, as well as security and confidentiality.

6.5.1 Intelligent dedicated search engines

Web search engines are software tools that help locate specific pieces of information on the WWW. Several such engines exist to date and are widely utilised by members of the Internet community. Some of the most popular include Alta Vista[19] developed by Digital Equipment Corporation, Yahoo![20] from Yahoo! Inc., Lycos[21] from Lycos, Inc. and HotBot[22] from HotWired, Inc. Each of these search engines has indexed a large portion of the documents available on the WWW, and their databases may be queried by keywords. The main drawback common to all such general purpose search engines is that their databases contain millions of documents and that queries often return too many unspecific results. For example, a search for *proteome* on the Alta Vista Web search engine returns 18,826 documents, while a query for *protein* produces more than 800,000 answers, most of which are probably irrelevant to the person who entered the query.

One remedy to this problem will consist of setting up dedicated search engines, which are query tools limited to a given field, thus restricting searches to potentially interesting areas. Several such search engines are already accessible on the Web. An example relevant to proteome research is SWISS-2DPAGE and Health On the Net's *2DHunt* (Boyer et al. 1996)[23] that provides links to documents and databases related to electrophoresis in general and 2-DE in particular (Fig. 6.16). A second remedy involves several ongoing research projects which are building increasingly intelligent search engines. These programs assist the user in formulating precise queries, filter the information based on acquired knowledge about the user's interests, and thus help him or her in obtaining only the data of importance (Etzioni and Weld 1994).

---

[19] http://altavista.digital.com/

[20] http://www.yahoo.com/

[21] http://www.lycos.com/

[22] http://www.hotbot.com/

[23] http://www.expasy.ch/ch2d/2DHunt/

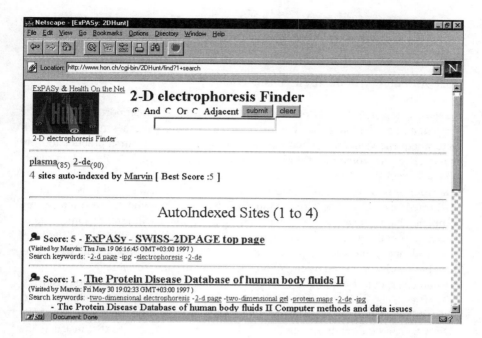

Fig. 6.16. The *2DHunt* Web search engine, designed to locate documents related to electrophoresis

6.5.2 Enhanced database integration

The tools and methods described in Sect. 6.3.3 and Sect. 6.3.4 let users access data from several databases that are interlinked by active cross-references. They allow easy navigation between databases over the Internet, but do not provide any means to integrate results of queries to multiple databases. For example, services such as SRS (see Sect. 6.2.3) perform multi-database searches, but local copies of all databases have to be maintained on the service's computer, thus making the checking of data integrity a difficult problem. New paradigms are currently under development that will provide the necessary tools to perform complex queries to databases distributed over the Internet.

One such concept is CORBA (Common Object Request Broker Architecture) developed by the Object Management Group (OMG) a consortium of software vendors and users (Siegel 1996). CORBA defines a standard mechanism by which objects transparently make requests to and receive responses from an Object Request Broker (ORB). ORBs that will play the role of an interface between clients and servers will be available on many computer platforms and will provide a common interface for client/server applications. In particular, it will allow for true database distribution over the net, as it will let any CORBA compliant application request data from other applications, independent of computer architecture, programming language or database management system.

6.5.3 Automation of database construction

As was stated in Chap. 3, protein identification plays an important role in the understanding of biological systems. More and more genomes are being completely sequenced and in the coming years it will become common practice to fully characterise the proteome of a given species, especially from 2-DE protein maps. Whilst current 2-DE analysis software provides the facilities to directly retrieve data from proteome databases (see Sect. 6.3.4), database construction is still being performed manually. The large amount of data that is going to be produced by complete characterisation of proteomes from 2-DE maps will require some kind of automation in the process of adding information to databases. With the rising level of confidence in the various protein identification tools, automatic data extraction from the results produced by these tools will soon become feasible.

Another aspect of automation lies at the level of protein identification itself. 2-DE software can be linked to a spot handling apparatus. Equipment such as matrix-assisted laser desorption / ionisation time of flight - mass spectrometers (MALDI-TOF MS) can be directly driven by 2-D image analysis software that instructs the equipment to perform its operations on the spots of interest, then sending mass spectrometric data directly to the various protein identification tools available on the WWW. Results can then be filtered and used to augment proteome databases.

6.5.4 Advanced data processing over the Internet

Most of current proteome-related data processing is carried out on the researcher's computer, either on his or her own data or on local copies of information obtained from one or more databases over the Internet. It is nevertheless questionable whether copying data over the net in order to be processed locally is efficient in terms of time and network load. Attempts to run programs on data that are located on remote computers have already been undertaken. For example, Lemkin (1997) proposes a tool on the WWW that allows the visual comparison of two 2-DE images retrieved from two different remotely located databases (Fig. 6.17).

This idea can certainly be generalised in order to process other remotely stored data without prior need to be downloaded. We may look at a fictional example. Let us assume we are trying to understand changes at the protein level in human Plasma of two populations, say disease A versus control. We have run three 2-DE gels of Plasma samples for each population. These six 2-DE images are analysed by a 2-DE analysis software package such as Melanie II. We know (or can find out easily) that Plasma master maps are available from 2-DE databases on the Internet. The six gel images are automatically compared with one of the master gels directly

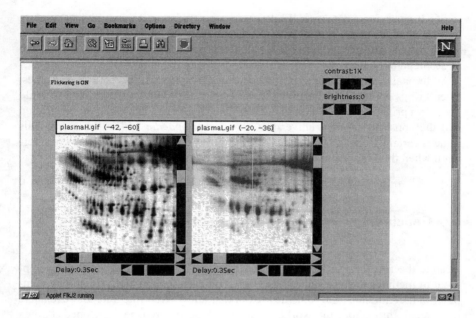

Fig. 6.17. The *flicker* program (Lemkin 1997) which visually compares 2-DE images over the Internet (here the Plasma master from SWISS-2DPAGE and a Plasma gel made by Carl Merril and coworkers at NIMH, Bethesda)

over the Internet, any differentially expressed proteins between the two populations are extracted, and data related to these proteins immediately retrieved from federated databases in order to collect all pertinent information about the proteins of interest. Fig. 6.18 (see p. 142) illustrates this fictional example, where six 2-DE images were compared to the Plasma map in SWISS-2DPAGE without being downloaded onto the local computer. Major variations in protein levels were extracted, and related data subsequently retrieved from the SWISS-2DPAGE and HSC-2DPAGE 2-DE databases, or from the SWISS-PROT protein sequence database and the OMIM on-line Mendelian Inheritance in Man database.

6.5.5 Security and confidentiality

A chapter on data transfer over the Internet would not be complete without mentioning security and confidentiality. Security is currently one of the main issues in Internet research and development. There is a general fear among Internet users that someone indiscreet might be "listening" to the net, and that he or she might find out what kind of data a person is retrieving. This is certainly a serious issue for electronic banking, for example, but pharmaceutical and biotechnology companies are also afraid that their trade secrets could be stolen by the competition when remotely querying a database. Whilst there is no evidence at this time that security holes will create serious problems to research laboratories or corporations, there is no doubt that advanced security tools such as cryptography will have to be put into widespread use before this issue is solved.

Additionally, the same question of confidentiality, when taken to its extremes, often brings companies to maintain local copies of public databases, and to build their own private database around it, thus hindering large scale database integration and the construction of a universal *Cyber-Encyclopaedia of the Proteome*. This issue is certainly outside the scope of this book, but it has to be taken into consideration when thinking about the future evolution of proteome databases.

6.6 Concluding remarks

Despite the fact that we have been working for over 5,000 years on the development of methods to store and retrieve data, and even though high technology computer and communication tools now exist that allow us to integrate and exchange knowledge throughout the world, nature has long beforehand developed its own system that is far more complex and much better performing than anything mankind has ever made. It is interesting to note that nature's data integration system, that is present in every organism, is exactly the subject of genome and proteome research, and that much work remains to be done before it is well understood. Also, many new concepts will have to be invented in computer science and related fields such as bioinformatics for us to efficiently deal with the large volume of proteomic data that will certainly be produced in the coming years. In particular, post-translational modifications of proteins induce numerous small shifts in the location of protein spots on two-dimensional electrophoresis, and we are only starting to perceive the computing challenge of building proteome databases that incorporate these variations. There is no doubt that the continuing progress being achieved in both proteome research and bioinformatics will contribute to further develop and implement the idea of a *Cyber-Encyclopaedia of the Proteome*.

Acknowledgments

The author would like to thank Marc Wilkins and Keith Williams for careful reading of the manuscript and helpful suggestions. Much of the work reported in this chapter could only be achieved thanks to the outstanding collaboration with Denis Hochstrasser, Amos Bairoch, Jean-Charles Sanchez and Manuel Peitsch, as well as the work of Vincent Baujard, Roberto Fabbretti, Elisabeth Gasteiger, Olivier

Golaz, Christine Hoogland, Patricia Palagi, Christian Pasquali, Florence Ravier, Luisa Tonella, Reynaldo Vargas and Daniel Walther.

References

Abola EE, Manning NO, Prilusky J, Stampf DR, Sussman JL (1996) The Protein Data Bank: current status and future challenges. J Res Natl Inst Stand Technol 101:231–241

Appel RD, Hochstrasser DF (1997) Computer analysis of 2-D images. In: Link AJ (ed) 2-D protein gel electrophoresis protocols. Humana Press, in press

Appel RD, Sanchez JC, Bairoch A, Golaz O, Miu M, Pasquali C, Vargas RJ, Hughes G, Hochstrasser DF (1993) SWISS-2DPAGE: a database of two-dimensional gel electrophoresis images. Electrophoresis 14:1232–1238

Appel RD, Bairoch A, Hochstrasser DF (1994) A new generation of information retrieval tools for biologists: the example of the expasy WWW server. Trends Biochem Sci 19:258–260

Appel RD, Bairoch A, Sanchez JC, Vargas JR, Golaz O, Pasquali C, Hochstrasser DF (1996a) Federated 2-DE database: a simple means of publishing 2-DE data. Electrophoresis 17:540–546

Appel RD, Sanchez JC, Bairoch A, Golaz O, Ravier F, Pasquali C, Hughes GJ, Hochstrasser D (1996b) The SWISS-2DPAGE database of two-dimensional polyacrylamide gel electrophoresis, its status in 1995. Nucleic Acids Res 24:180–181

Appel RD, Bairoch A, Hochstrasser DF (1997) Constructing a 2-D database for the WWW. In: Link AJ (ed) 2-D protein gel electrophoresis protocols. Humana Press, in press

Bairoch A, Apweiler R (1997) The SWISS-PROT protein sequence data bank and its supplement TrEMBL. Nucleic Acids Res 25:31–36

Bairoch A, Bucher P, Hofmann K (1997) The PROSITE database, its status in 1997. Nucleic Acids Res 25:217–222

Benson DA, Boguski M, Lipman DJ, Ostell J (1996) GenBank. Nucleic Acids Res 24:1–5

Berners-Lee TJ, Cailliau R, Groff JF, Pollermann B (1992) World-Wide Web: The information universe. Electronic Networking: Research, Applications and Policy 2:52–58

Boyer C, Baujard O, Baujard V, Aurel S, Selby M, Appel RD (1996) HON automated database of health and medical information. In: MEDNET96 — European Congress on the Internet in Medicine, Brighton, U.K.,14–17 October 1996, pp 36–37

Comer DE (1988) Internetworking with TCP/IP: principles, protocols, and architecture. Prentice Hall, Inc., Englewood Cliffs, NJ

Corbett JM, Wheeler CH, Baker CS, Yacoub MH, Dunn MJ (1994) The human myocardial two-dimensional gel protein database: update 1994. Electrophoresis 15:1459–1465

Dayhoff, MO (1978) Atlas of protein sequence and structure, vol. 5 supp. 1–3. National Biomedical Research Foundation, Washington D.C.

Devereux J, Haeberli P, Smithies O (1984) A comprehensive set of sequence analysis programs for the VAX. Nucleic Acids Res 12:387–395

Dimant D, Rappaport U (eds) (1992) The Dead Sea Scrolls: forty years of research. E. J. Brill, Leiden

Etzioni O, Weld D (1994) A softbot-based interface to the internet. Comm of the ACM 37:72–76

Etzold T, Argos P (1993) SRS: an indexing and retrieval tool for flat file data libraries. Comput Appl Biosci 9:49–57

George DG, Barker WC, Mewes HW, Pfeiffer F, Tsugita A (1996) The PIR-International protein sequence database. Nucleic Acids Res 24:17–20

Golaz O, Hughes GJ, Frutiger S, Paquet N, Bairoch A, Pasquali C, Sanchez JC, Tissot JD, Appel RD, Walzer C, Balant L, Hochstrasser DF (1993) Plasma and red blood cell protein maps: The 1993 update. Electrophoresis 14:1232–1238

Harger C, Skupski M, Allen E, Clark C, Crowley D, Dickinson E, Easley D, Espinosa-Lujan A, Farmer A, Fields C, Flores L, Harris L, Keen G, Manning M, McLeod M, O'Neill J, Pumilia M, Reinert R, Rider D, Rohrlich J, Romero Y, Schwertfeger J, Seluja G, Siepel A, Schad PA (1997) The Genome Sequence DataBase version 1.0 (GSDB): from low pass sequences to complete genomes. Nucleic Acids Res 25:18–23

Harper R (1994) Access to DNA and protein databases in the internet. Curr Opin Biotechnol 5:4–18

Henikoff S (1993) Sequence analysis by electronic mail server. Trends Biochem Sci 18:267–268

Hochstrasser DF, Appel RD, Golaz O, Pasquali C, Sanchez JC, Bairoch A (1995) Sharing of worldwide spread knowledge using hypermedia facilities & fast communications protocols (Mosaic & World-Wide Web): The example of ExPASy. Methods Inf Med 34:75–78

Hoogland C, Baujard V, Sanchez JC, Hochstrasser DF, Appel RD (1997) Make2ddb: a simple package to set up a 2-DE database on the WWW. Electrophoresis 18, in press

Lemkin PF (1997) Comparing two-dimensional gels across the Internet. Electrophoresis 18:461–470

Nissen HJ, Damerow P, Englund RK (1990) Frühe Schrift und Techniken der Wirtschaftsverwaltung im alten vorderen Orient: Informationsspeicherung und -verarbeitung vor 5000 Jahren. Franzbecker, Bad Salzdetfurth, pp 11–43

Patton WF, Lim MJ, Shepro D (1997) Image acquisition in 2-D electrophoresis. In: Link AJ (ed) 2-D protein gel electrophoresis protocols. Humana Press, in press

Pearson P, Francomano C, Foster P, Bocchini C, Li P, McKusick V (1994) The status of online Mendelian inheritance in man (OMIM) medio 1994. Nucleic Acids Res 22:3470–3473

Peitsch, MC (1996) ProMod and Swiss-Model: Internet-based tools for automated comparative protein modelling. Biochem Soc Trans 24:274–279.

Peitsch MC, Stampf DR, Wells TNC, Sussman JL (1995) The Swiss-3DImage collection and PDB-Browser on the World-Wide Web. Trends Biochem Sci 20:82–84

Rodriguez-Tome P (1997) Searching the dbEST database. Methods Mol Biol 69:269–283

Rodriguez-Tome P, Stoehr PJ, Cameron GN, Flores TP (1996) The European Bioinformatics Institute (EBI) databases. Nucleic Acids Res 24:6–12

Schuler GD, Epstein JA, Ohkawa H, Kans JA (1996) Entrez: molecular biology database and retrieval system. Methods Enzymol 266:141–162

Siegel J (ed) (1996) CORBA Fundamentals and Programming. John Wiley and Sons, New York

Sonnhammer EL, Kahn D (1994) Modular arrangement of proteins as inferred from analysis of homology. Protein Sci 3:482–492

Wilbur WJ, Yang Y (1996) An analysis of statistical term strength and its use in the indexing and retrieval of molecular biology texts. Comput Biol Med 26:209–222

Wilkins MR, Sanchez JC, Bairoch A, Hochstrasser DF, Appel RD (1996) Proteome Projects: integrating two-dimensional gel databases using the Melanie II software. Trends Biochem Sci 21:496–497

Wilkins MR, Lindskog I, Gasteiger E, Bairoch A, Sanchez JC, Hochstrasser DF, Appel RD (1997a) Detailed peptide characterization using PeptideMass — a World-Wide Web accessible tool. Electrophoresis 18:403–408

Wilkins MR, Gasteiger E, Bairoch A, Sanchez JC, Williams KL, Appel RD, Hochstrasser DF (1997b) Protein identification and analysis tools in the ExPASy server. In: Link AJ (ed) 2-D protein gel electrophoresis protocols. Humana Press, in press

7 Large-Scale Comparative Protein Modelling

Manuel C. Peitsch and Nicolas Guex

7.1 Introduction

Proteins are the working molecules in most biological processes and a large portion of experimental biology is aimed at understanding their function. To understand function it is necessary to know the shape of the protein and where or how it interacts with other proteins or substrates. This is approached by first solving the three-dimensional (3-D) structure and then changing the protein sequence to see how the structure and/or function is altered. Site-directed mutagenesis experiments are generally designed to provide information about the involvement of a protein's specific residues in enzyme-based reactions, molecular recognition events, protein stability and during drug design projects. The experimental elucidation of 3-D structures by X-ray crystallography or NMR is, however, often hampered by difficulties in obtaining sufficient pure protein, diffracting crystals, the size of the protein or its subunits for NMR studies and many other technical aspects. This is illustrated by the fact that the SWISS-PROT/TrEMBL database (Bairoch and Apweiler 1997) contains around 175,000 sequences, while the Brookhaven Protein Data Bank (PDB) (Abola et al. 1996) contains about 6,000 3-D structures from 1,700 different proteins. The number of solved 3-D structures is increasing very slowly compared to the rate of sequencing of novel cDNAs, and no structural information is available for the vast majority of known protein sequences. This gap will, of course, further increase as genome sequencing projects continue to yield large numbers of novel sequences.

In this context it is not surprising that methods for predicting 3-D protein structures have gained a lot of interest. The most reliable of these methods, comparative protein modelling, is based on the observation that proteins with similar sequences have similar structures (Chothia and Lesk 1986). It extrapolates the structure of a new protein sequence from the known 3-D structure of related family members (for review see Bajorath et al. 1993).

Protein modelling not only requires expensive computer hardware and software, but also expert knowledge of how to manipulate structural information. Therefore for many years, only a limited number of scientists had access to these tools. In order to simplify this process, we have established the SWISS-MODEL server for automated comparative protein modelling (Peitsch 1995; Table 7.1), and the Swiss-

Table 7.1. Internet locations for 3-D structure and related resources (see also Table 5.1)

Database or Service	Internet URL Address
ExPASy molecular biology server	http://www.expasy.ch/
SWISS-MODEL	http://www.expasy.ch/swissmod/SWISS-MODEL.html
SWISS-MODEL Repository	http://www.expasy.ch/swissmod /swmr-top.html
Swiss-PdbViewer	http://www.expasy.ch/spdbv/mainpage.htm
	http://www.pdb.bnl.gov/expasy/spdbv/mainpage.htm
SWISS-3DIMAGE	http://www.expasy.ch/sw3d/sw3d-top.html
SWISS-PROT	http://www.expasy.ch/sprot/sprot-top.html
SWISS-2DPAGE	http://www.expasy.ch /ch2d/ch2d-top.html
WebMol (Dirk Walther)	http://www.embl-heidelberg.de/cgi/viewer.pl

PdbViewer (Guex and Peitsch 1996) which is a sequence-to-structure workbench integrated with the server. More recently we have used the software framework of the server to build large collections of protein models, which can now be obtained from the SWISS-MODEL Repository, a new database for theoretical protein models (Peitsch et al. 1997; Peitsch 1997).

7.2 Methods and programs

7.2.1 Identification of modelling templates

Automated comparative protein modelling requires at least one sequence of known 3-D structure with significant similarity to the target sequence. In order to determine if a modelling request can be carried out, the SWISS-MODEL server compares the target sequence with a database of sequences derived from the Brookhaven Protein Data Bank (PDB), using both FastA (Pearson and Lipman 1988) and BLAST (Altschul et al. 1990) sequence comparison programs. Sequences with a FastA score 10.0 standard deviations above the mean of the random scores and a BLAST Poisson unlikelihood probability $P(N)$ lower than 10^{-5} are considered for the model building procedure. The choice of template structures is further restricted to those which share at least 35% residue identity with 40% of the target sequence as determined by the sequence alignment algorithm SIM (Huang and Miller 1991).

The above procedure might allow the selection of several suitable templates for a given target sequence, and up to ten templates can be used during the modelling process. The best template structure — the one with the highest sequence similarity to the target — serves as the *reference*. All other selected templates will be superposed onto it in 3-D. The 3-D match is carried out by superposing corresponding Cα atom pairs (i.e. the carbon atoms in the peptide backbone that carry the amino acid R groups) selected automatically from the highest scoring local sequence alignment determined by SIM. This superposition is then optimised by maximising the number of Cα pairs in the common core (Chothia and Lesk 1986) while minimising their relative mean square (rms) deviation. Each residue of the reference structure is then aligned with a residue from every other available template structure if their Cα atoms are located within 3.0 Å. This generates a structurally corrected multiple sequence alignment.

7.2.2 Aligning the target sequence with the template sequence(s)

The target sequence is aligned with the template sequence or, if several templates were selected, with the structurally corrected multiple sequence alignment using the best-scoring diagonals obtained by SIM. Residues which should not be used for model building, for example those located in non-conserved loops, are ignored during the modelling process. Thus, the common core of the target protein and the loops completely defined by at least one supplied template structure will be built.

7.2.3 Building the model

The coordinates of the model are built using ProMod (Peitsch 1996). The (multiple) sequence alignment serves as a correspondence table between target sequence and template structures from which a weighted average structural framework is derived. This framework is then completed by the addition of lacking or incomplete loop structures (which correspond to insertion and deletions in the sequence alignment), the rebuilding of undefined backbone atoms, the correction of ill-defined side chain geometries and the addition of lacking side-chains (Bajorath et al. 1993; Peitsch 1996).

7.2.4 Model refinement

The final step of the coordinate generation process is the idealisation of the model stereochemistry, and consists mainly of the optimisation of bond geometry and the removal of unfavourable non-bonded contacts. This step is performed by energy minimisation using the CHARMm force field (Brooks et al. 1983).

Excessive energy minimisation will cause the model to deviate markedly from the original model, which is not suitable and should be avoided. Indeed, experience has shown that the changes induced by force field computations do not improve the accuracy of the model with respect to a control experimental structure. Thus, we use as few as possible cycles of minimisation, but sufficient to improve the stereochemistry of the model.

7.2.5 Assessing the model quality

The quality of a model is determined by two distinct criteria, which will determine its applicability (see Sect. 7.2.6). First, the correctness of a model is dictated by the quality of the sequence alignment used to guide the modelling process. If the sequence alignment is wrong in some regions, then the spatial arrangement of the residues in this portion of the model will be incorrect. Second, the accuracy of a model is essentially limited by the deviation of the used template structure(s) relative to the experimental control structure. This limitation is inherent to the methods used, since the models result from an extrapolation. As a consequence, the core Cα atoms of protein models which share 35 to 50% sequence identity with their templates will generally deviate by 1.0 to 1.5 Å from their experimental counterparts, as do experimentally elucidated structures (Chothia and Lesk 1986). It may be noted that deviation of less than 1.0 Å is considered to be a very good fit. One should, however, not overlook the contributions of the templates to the model accuracy. The templates, which are obtained through experimental approaches, are subject to structural variations not only caused by experimental errors and differences in data collection conditions (such as temperature, Tilton et al. 1992), but also because of different crystal lattice contacts and the presence or absence of ligands.

This can be illustrated with an example. The structure of interleukin-4 (IL-4) (Harrison et al. 1995 and references therein), a cytokine consisting of a 130 residue four helix bundle, was elucidated by X-ray crystallography as well as by NMR. The backbones of three IL-4 crystal structures (PDB entries 1RCB, 2INT and 1HIK) show rms deviations of 0.4 to 0.9 Å, while those of three IL-4 NMR forms (PDB entries 1ITM, 1CYL and 2CYK) deviated by 1.2 to 2.6 Å. These values illustrate the structural differences due to experimental procedures and the molecular environment at the time of data collection. Therefore:

"a protein model derived by comparative methods cannot be more accurate than the difference between the NMR and crystallographic structure of the same protein" (Harrison et al. 1995).

Almost every protein model contains non-conserved loops which are expected to be its least reliable parts. Indeed, non-conserved loops often deviate markedly from experimentally determined control structures. In many cases, however, these loops also correspond to the most flexible parts of the structure, as evidenced by their high crystallographic temperature factors or multiple solutions in NMR experiments. On the other hand, the core residues — that are the least variable in any given protein family — are usually found in essentially the same orientation as in experimental control structures, while far larger deviations are observed for surface amino acids. This is expected since the core residues are generally well conserved and the rotamers of their side chains are constrained by neighbouring residues. In contrast, the more variable surface amino acids will tend to show more deviations since there are fewer steric constraints imposed upon them.

Some structural aspects of a protein model can be verified using methods based on the inverse folding approach. Two of them, namely the 3-D profile based verification method (Lüthy et al. 1992) and ProsaII (Sippl 1993), are widely used. The 3-D profile of a protein structure is calculated by adding the probability of occurrence for each residue in its 3-D context (Lüthy et al. 1992). Each of the twenty amino acids has a certain probability to be located in one of the 18 environmental classes (defined by criteria such as solvent-accessible surface, buried polar and exposed non-polar area and secondary structure). In contrast, ProsaII relies on empirical pseudo-conformational energy potentials derived from the pair-wise interactions observed in well-defined protein structures. These terms are summed over all residues in a model and result in a more (more negative) or less (more positive) favourable energy. Both methods can detect a global sequence-to-structure incompatibility and errors corresponding to topological differences between template and target. They also allow the detection of more localised errors such as β-strands that are "out of register" or buried charged residues. The two inverse folding approaches, however, are unable to detect the more subtle structural inconsistencies often localised in non-conserved loops, and cannot provide an assessment of the correctness of their geometry.

7.2.6 Application of protein models

Protein models obtained by comparative modelling methods can be classified into three broad categories. i) Models which are based on incorrect alignments between target and template sequences. Such alignment errors, which generally reside in the inaccurate positioning of insertions and deletions, are caused by the weaknesses of the alignment algorithms and can often not be resolved in the absence of a control experimental structure. It is, however, often possible to correct such errors by producing several models based on alignment variants and by selecting the most "sensible" solution. Nevertheless, it turns out that such models are often useful as the errors are not located in the area of interest, such as within a well-conserved active site. ii) Models based on correct alignments are of course much better, but they can still be of medium to low accuracy as the templates used during the modelling process have a medium to low sequence similarity with the target sequence. Such models, as the ones described above, are very useful tools for designing mutagenesis experiments aimed at identifying key structural elements. They may, however, not be of great assistance for detailed ligand binding studies. iii) The last category of models comprises all those which were built based on templates that share a high degree of sequence identity (> 70%) with the target. Such models have proven useful during drug design projects and allowed key decisions to be taken for compound optimisation and chemical synthesis. For instance, models of several species variants of a given enzyme can guide the design of more specific non-natural inhibitors.

However, having said the above, there are times when a model is expected to be useful but is not, and conversely, there are occasions when at first sight a model has failed but on closer inspection has in fact been effective.

7.2.7 Sequence-to-structure workbench

One limiting step in protein modelling is the availability of powerful graphical interfaces to visualise and analyse molecular models on low-cost computers. To overcome this limitation we have developed the Swiss-PdbViewer for Macintosh and PC (Fig. 7.1, see p. 143; Table 7.1; Guex and Peitsch 1996). Its main strength as a molecular viewing software is its powerful user-oriented interface which provides easy-to-use selection, display, measurement and comparison tools. Furthermore, the Swiss-PdbViewer is a sequence-to-structure workbench which acts as a front end to SWISS-MODEL and eases the process of protein modelling by providing an integrated work environment. Using a World-Wide Web browser as a helper application, one can search for the most suitable modelling template(s) and download the corresponding coordinate file(s). Integrated sequence alignment tools and structure superposition algorithms allow the mapping of a new sequence onto a given modelling template. The automatically generated sequence alignment can be optimised by the user to override any errors made by the sequence alignment algorithm. The mapped model is updated in real time following each change in the sequence alignment, which provides immediate feedback on every alteration. Once the sequence alignment is satisfactory, the corresponding request can be submitted to SWISS-MODEL.

The coordinate files returned by SWISS-MODEL can be readily loaded into the Swiss-PdbViewer and analysed. For instance, they can be coloured according to their C-factor (Peitsch 1996), a factor computed by SWISS-MODEL describing the relative confidence one can have in each residue's position.

7.3 Modelling of Proteomes

7.3.1 Large-scale protein modelling

There is no doubt that the conclusions drawn from the scrutiny of multiple sequence alignments are far more reliable than those derived from the analysis of an isolated sequence. The mere identification of conserved and variable residues already provides useful anchor points for site-directed mutagenesis experiments aimed at understanding the function of a protein. Likewise, comparative structure analysis involving several members of a protein family, as opposed to single structure inspection, is expected to be invaluable for the understanding of functional differences and for rational drug design.

Within the last two years, the genome sequences of several bacteria and yeast have been completely determined. Even more genomes will be sequenced soon. Therefore, ever-increasing collections of truly orthologous and paralogous sequences are becoming available. Using large-scale protein modelling methods, it is possible to rapidly generate model structures for all members of a protein family if at least one of its members has a known 3-D structure. These models can then be superposed in 3-D space and comparative structure analysis undertaken to understand the differences observed between, for example, a certain protein derived from different species.

A comparison of all entries of SWISS-PROT (release 34) with a non-redundant subset of all sequences with a known 3-D structure revealed that approximately 15% of the proteins of any organism belong to a family with a suitable modelling template. This proportion has increased by approximately 5% over the last year, which allows us to speculate that we might be able to derive comparative models for over 30% of all proteins at the beginning of the next millennium.

In order to initiate the process of large-scale protein modelling (Peitsch et al. 1997; Peitsch 1997), we have taken a species-based approach, and submitted — in batch mode — all known proteins from the proteomes of *Escherichia coli, Haemophilus influenzae, Mycoplasma genitalium, Mycobacterium tuberculosis, Bacillus subtilis* and *Saccharomyces cerevisiae* to the SWISS-MODEL server. In this fully automated approach we required at least 35% residue identity within the segments aligned to the modelling template. This ensures that the quality of the resulting models does not fall below an acceptable threshold; specifically, a common core at 1.5 Å rms deviation. The protein models created with this method were placed in the SWISS-MODEL Repository. The number of models obtained for each species is shown in Table 7.2.

Table 7.2. Number of protein models generated for selected organisms

Species	Submitted[a]	Models[b]
Bacillus subtilis	1783	263 (14.8%)
Escherichia coli	3614	519 (14.3%)
Haemophilus influenzae	1591	236 (14.8%)
Mycobacterium tuberculosis	473	78 (16.5%)
Mycoplasma genitalium	433	65 (15.0%)
Saccharomyces cerevisiae	6615[c]	933 (14.1%)

[a]number of sequences in SWISS-PROT/TrEMBL submitted to SWISS-MODEL
[b]number of protein models obtained (in parentheses: percent of submitted)
[c]the number of sequences here is greater than the number of open reading frames predicted from the genome because of redundancy in the SWISS-PROT/TrEMBL databases.

7.3.2 Model Repository

All protein models generated automatically by SWISS-MODEL are annotated with information regarding the template(s) used for model building. The sequence alignment between the template and the model (target) is also provided in a numerical format. In order to ease the connectivity with other databases, each model has the same identification and accession codes as the corresponding SWISS-PROT entry. The models are stored as individual files and can be accessed through the SWISS-MODEL Web pages (Table 7.1). In addition, every time a SWISS-PROT or a SWISS-2DPAGE entry is requested from the ExPASy Web server (Table 7.1; Appel et al. 1994), a hypertext link to the coordinates file is provided if a corresponding model exists in the SWISS-MODEL Repository (Peitsch et al. 1997; Peitsch 1997). Model coordinates can be readily downloaded and imported into the sequence-to-structure workbench Swiss-PdbViewer (Guex and Peitsch 1996), or into a viewing software such as RasMol (Sayle and Milner-White 1995). Alternatively one can use the WebMol Java applet written by Dirk Walther (Table 7.1).

7.4 Examples

7.4.1 Obtaining a 3-D structure from the SWISS-MODEL Repository

The entries in the SWISS-MODEL Repository are annotated with the same definition as their SWISS-PROT counterparts. It is therefore possible to search an index to find models of interest. The SWISS-MODEL Repository and the server's top page provide links to a search tool which allows users to submit simple keyword queries as well as SWISS-PROT AC or ID codes (see Sect. 5.2.1). The corresponding model(s), if any, can then be downloaded directly into the Swiss-PdbViewer or saved to disk. This process is illustrated in Fig. 7.2 (see p. 144 and 145).

7.4.2 Modelling the Fas-ligand protein using the Swiss-PdbViewer and SWISS-MODEL

This sample project makes heavy use of the World-Wide Web and all searches and queries were submitted to the ExPASy server (Table 7.1). Initially, the SWISS-PROT database was searched with the keyword FASL (to represent the Fas-ligand). From the list of entries found, we selected the entry FASL_HUMAN and saved the content of that page to disk in HTML format, to retain all hyperlinks (see 6.3.2). This file was then loaded into the Swiss-PdbViewer from the SWISS-MODEL pull-down menu. Raw sequences, devoid of atomic coordinates, appeared as a single helix on the screen. From the same menu, it was then possible to search for suitable modelling templates using Netscape or other browsers as helper applications. The query form was automatically completed by the software and its content was submitted to perform a sequence similarity search with BLAST (Altschul et al. 1990). A selection of best-scoring sequences of known 3-D structure was sent back by ExPASy, and each selected template (the first chain of 1TNF and the second chain of 1TNR) was downloaded into the Swiss-PdbViewer. Both templates were superposed in 3-D space and a structure-based sequence alignment was generated. The sequence of FASL_HUMAN was aligned with the template sequence and the sequence alignment corrected by manual intervention in order to correct a misplaced deletion. Once the sequence alignment was satisfactory, we submitted the modelling request to SWISS-MODEL from the corresponding menu. Again, an HTML form was automatically generated by the Swiss-PdbViewer and the request submitted via Netscape. The resulting model was sent back within 20 minutes and its coordinates could be analysed (Fig. 7.3, see p. 146). The above example is further detailed in the tutorial pages for Swiss-PdbViewer on the ExPASy server (Table 7.1) and can be followed step-by-step.

7.5 Concluding remarks

In this chapter we have shown that comparative protein modelling methods can be made available to a wide community, and that they can be used successfully to predict the structure of large sets of proteins through automated approaches. One should however not forget that comparative protein modelling still requires at least one suitable template structure, which can only be obtained by experimental means. It is clear that the experimental structure elucidation methods have made amazing progress and that the number of comparative models which one can derive will increase in the near future.

Comparative protein modelling however still suffers from a number of limitations, which will require further research and the development of novel approaches. For instance, one can presently not model large multi-domain complexes, mainly because no suitable modelling templates are available and because the prediction of protein-protein contacts is inaccurate and still in its infancy. Another challenge is that one cannot easily predict the influence of ligand binding and post-translational modifications on the protein structure. This comment applies to all structures, whether they are obtained through experimental or theoretical approaches, although there are a number of experimentally determined structures that include both ligand-bound and glycosylated forms of proteins. It can be expected that such experimentally derived structures will ultimately assist modelling in this interesting area of structural biology.

Finally, we hope that by developing a freely available integrated protein modelling environment and a model database, we will allow an ever-increasing number of molecular biologists to use these methods to plan experiments.

References

Abola EE, Manning NO, Prilusky J, Stampf DR, Sussman JL (1996) The Protein Data Bank: current status and future challenges. J Res Natl Inst Stand Technol 101:231–241

Altschul SF, Gish W, Miller W, Myers EW, Lipman DJ (1990) Basic local alignment search tool. J Mol Biol 215:403–410

Appel RD, Bairoch A, Hochstrasser DF (1994) A new generation of information retrieval tools for biologists: the example of the ExPASy WWW server. Trends Biochem Sci 19:258–260

Bairoch A, Apweiler, R (1997) The SWISS-PROT protein sequence data bank and its new supplement TrEMBL. Nucleic Acids Res 25:31–36

Bajorath J, Stenkamp R, Aruffo A (1993) Knowledge-based model building of proteins: concepts and examples. Prot Sci 2:1798–1810

Brooks BR, Bruccoleri RE, Olafson BD, States DJ, Swaminathan S, Karplus M (1983) CHARMM: a program for macromolecular energy, minimization and dynamics calculation. J Comp Chem 4:187–217

Chothia C, Lesk AM (1986) The relation between the divergence of sequence and structure in proteins. EMBO J 5:823–826

Guex N, Peitsch MC (1996) Swiss-PdbViewer: a fast and easy-to-use PDB viewer for Macintosh and PC. Protein Data Bank Quarterly Newsletter 77:7

Harrison RW, Chatterjee D, Weber IT (1995) Analysis of six protein structures predicted by comparative modeling techniques. Proteins Struct Func Genet 23:463–471

Huang X, Miller M (1991) A time-efficient, linear-space local similarity algorithm. Adv Appl Math 12:337–357

Lüthy R, Bowie JU, Eisenberg D (1992) Assessment of protein models with three-dimensional profiles. Nature 356:83–85

Pearson WR, Lipman DJ (1988) Improved tools for biological sequence comparison. Proc Natl Acad Sci USA 85:2444–2448

Peitsch MC (1995) Protein modelling by E-Mail. Bio/Technology 13:658–660

Peitsch MC (1996) ProMod and Swiss-Model: Internet-based tools for automated comparative protein modelling. Biochem Soc Trans 24:274–279

Peitsch MC (1997) Large scale protein modelling and model repository. In: Gaasterland T, Karp P, Karplus K, Ouzounis C, Sander C and Valencia A (eds) Proceedings of the fifth international conference on intelligent systems for molecular biology (vol 5) AAAI Press, pp 234–236

Peitsch MC, Wilkins MR, Tonella L, Sanchez JC, Appel RD, Hochstrasser DF (1997) Large scale protein modelling and integration with the SWISS-PROT and SWISS-2DPAGE databases: the example of *Escherichia coli*. Electrophoresis 18:498–501

Sayle RA, Milner-White EJ (1995) RASMOL: biomolecular graphics for all. Trends Biochem Sci 20:374–376

Sippl MJ (1993) Recognition of errors in three-dimensional structures of proteins. Proteins Struct Funct Genet 17:355–362

Tilton RF Jr, Dewan JC, Petsko GA (1992) Effects of temperature on protein structure and dynamics: x-ray crystallographic studies of the protein ribonuclease-A at nine different temperatures from 98 to 320 K. Biochemistry 31:2469–2481

8 Clinical and Biomedical Applications of Proteomics

Denis F. Hochstrasser

8.1 Introduction

The ultimate goal in medicine is to prevent disease, or cure disease when it appears but before any permanent damage has taken place or side effects become evident. Treatment of disease, if not curative, should provide palliative care and relieve the symptoms. The selection of preventive action or treatment choice rely on clear recognition of the precise disease status of the patient. Diagnostic, from the Greek word "diagnosticos", means capable of recognition. A diagnostic procedure helps to establish a patient condition or diagnosis and to classify patients into categories. Patient diagnosis has to date been established through physicians taking patient histories and doing physical examinations, and then if needed through analyses such as blood tests and more sophisticated radiological procedures. Once a diagnosis has been established, the prognosis, from the Latin "prognosticus" meaning knowing in advance, must be defined to orientate the treatment decision.

Any procedure in medicine, regardless of the cost constraint, can be justified only if it has an impact on: a) the patient diagnosis, in other words on categorising patients into classes, or b) the patient prognosis, helping to know in advance what will happen, and c) on the treatment definition or selection. However, a procedure can also be justified: d) if it has a good positive or negative predictive value in medicine, and e) associated with d) if and only if preventive measures exist and can be taken. In most cases, it is unethical to predict disease when no preventive measure or treatment exists.

To group patients into categories is getting more and more difficult as our diagnostic tools become more and more refined. Indeed when one considers only a patient's fingerprints, every patient is unique. Now, regardless of today's interpretation difficulties, our diagnostic tests have to be very sensitive to exclude diseases and/or very specific to confirm diagnoses. The prognostic knowledge by diagnostic categories is nowadays insufficient and tailor-made prognostic evaluation is increasingly required for each patient. For example, the metastatic potential of a patient tumour is most likely unique for a given patient. It evolves during time for that patient and can be quite different from one patient to the next even if patients appear to have identical clinicopathological findings.

To achieve the ultimate goals of medicine, we need a better understanding of disease processes and evolution. This chapter will explore how the proteome approach can be used to analyse patient samples in fundamental and applied medical research, and will also discuss how this contributes to our understanding of disease.

8.2 Biotechnology, genomics and medicine to date

Biotechnology has great potential to improve or enhance medical care. It enables improved diagnostic measures and tests, leads to improved prognostic evaluation, facilitates engineering new and/or better therapeutics, helps establishing predictive methods and offers preventive measures. All of the above should at all times be driven by ethics.

For some time, a driving force in biotechnology has been genomics. Numerous scientists have turned their effort towards understanding genetics, gene sequences and gene regulation. These tremendous developments have undoubtedly made a major revolution in clinical medicine, especially in unravelling genetic disease processes. But in the clinical laboratory the impact so far has been mainly in the domain of infectious diseases, helping to detect infections agents early by Polymerase Chain Reaction (PCR) or Reverse Transcriptase Polymerase Chain Reaction (RT-PCR), and in the domain of haematology and oncology where, for example, fluorescent *in situ* hybridisation (FISH) can help to better characterise the oncogenic process in cells.

One explanation for the somewhat limited impact of genetics in applied laboratory medicine so far is the following:

"The range of human phenotypes/diseases for which our burgeoning bio-molecular data base is sufficient to provide understanding, diagnosis, and therapy is small. Only 2 percent of our total disease load is related to monogenic causality, and even here the final phenotype is modulated by many factors" (Strohman 1994; also see Strohman 1993).

As further stated by Strohman, genetic analysis in itself cannot today predict or diagnose multigenic diseases. Complex gene interactions, cellular events or environmental influences modify gene expression and/or protein post-translational modifications. Several examples of epigenetic regulation in cells and in physiological systems or pathophysiological conditions demonstrate the role of redundant genes and metabolic and other cellular networks. Numerous overlapping pathways accomplish similar tasks and adapt themselves to achieve similar endpoints. In consequence, gene defects may be part of, but are not always sufficient to predict disease or diagnose illness. The prognosis of disease or choice of treatment may entirely depend on epigenetic regulation or dysregulation and must take into account the patient's environment.

Redundancy of functional information in cell regulation has been shown by several authors (Brenner et al. 1990). The statement one-gene-one-unique-effect is no longer tenable. Only in monogenic disease does the dogma of DNA to RNA to protein to phenotype or disease apply. But, as stated above, monogenic diseases seem to account for less than 2% of the total disease load and even here there are strong environmental and genetic background effects. For example, epigenetic regulation explains why, in numerous gene knockout experiments, cells or organisms display normal phenotypes despite gene deletion. In immunology and cytokine research, it is well established that environmental factors modulate a variety of cellular activities (Nathan and Sporn 1991).

Strong environmental factors influence many forms of disease such as cardiovascular diseases or cancer development and aggressiveness. For example, the raw fish diet and lack of vitamin C have been linked to the high incidence of stomach cancer in Japanese and Norwegian populations. Thus we can anticipate that even if the expression of thousands of genes can be catalogued at once with chips such as those developed by Affymetrics™, the lack of correlation between mRNA abundance and protein level will undermine this simultaneous gene detection method in multigenic diseases.

On a more positive note, numerous gene products have been recombinantly expressed, including insulin, granulocyte colony stimulating factor (G-CSF), erythropoietin and the recombinant tissue plasminogen activator (rTPA). They are now very successfully used routinely in the clinical world in several well defined illnesses. Genetic therapy offers very exciting potential applications in a few hereditary diseases such as cystic fibrosis, diabetes and even in cancer treatment.

8.3 The application of proteomics to medicine

As shown above, the complexity of interactions between the environment, genes and their products is tremendous. Therefore an approach complementary to genomics is required in clinical situations to better understand epigenetic regulation and get closer to a "holistic" medical approach.

In clinical medicine, the patient history and physical examination have always been essential to understand the patient condition and in most cases to establish a diagnosis. As stated in the article entitled "Epigenesis and complexity: the coming Kuhnian revolution in biology":

"*Also to be considered in assessing a mutant gene effect on a complex disease phenotype is the unique life history of the individual in question (diet, age, physical activity, stress, etc.). Therefore, understanding of complex function may in fact be impossible without recourse to influences outside of the genome*" (Strohman 1997; also see Strohman 1995).

The potential clinical applications of 2-D PAGE have previously been reviewed (Hochstrasser and Tissot 1993; Young and Tracy 1995). The application of proteomics to the analysis of body fluids and tissue biopsies is particularly useful. Specific applications include: a) identifying the origin of body fluid samples (whether they are spinal, cystic, serum, pleural, ascitic, etc.) or the origin of a tissue biopsy, b) analysing protein phenotypes and protein post-translational modifications (e.g. apolipoprotein E and J, haptoglobin) in fluid, cells or tissues, c) examining the clonality of immunoglobulins and detecting clones which are not seen with conventional techniques (in genetic, autoimmune, infectious or neoplasic disorders such as multiple sclerosis, haemolytic anaemia, Lyme's disease, CMV-infection, B-hepatitis, Sjögren's syndrome or lyphoproliferative disorders), d) monitoring disease processes and protein expression (in inflammation, early renal or hepatic insufficiency, nutrition disorders, toxicology, etc.), and e) discovering new disease markers and/or patterns in body fluids, cells or tissue biopsies.

But one should keep in mind the following words of caution:

"...In the absence of simple systems to identify and quantify individual proteins or groups of proteins it is unlikely that clinical applications will increase... Cost-containment pressures within the clinical laboratory will prevent the technique from becoming widely used in the clinical laboratory until it can clearly demonstrate that it can produce clinically important and necessary information that can not be obtained by other means..." (Young and Tracy 1995).

The goal of a proteomic approach is the massively parallel analysis of expressed proteins. New technologies will undoubtedly enhance the display aspects of 2-D PAGE and the protein characterisation potential, giving insights similar to those provided by the genomic approach. Eventually this global view could make a series of clinical tests redundant and therefore decrease the sample evaluation cost. As described in the introduction, this approach should be more sensitive and/or specific than simpler conventional techniques for routine diagnostic, prognostic or therapeutic evaluation of patient samples, and also should generally decrease the cost/benefit ratio in medical terms of the patient investigation and treatment follow-up.

8.4 Disease diagnosis from body fluids

The potential impact of a proteomic approach on patient care can be illustrated with the history of several patients who suffered either meningitis following head trauma, acute respiratory distress, renal insufficiency, haemolytic anaemia or dementia (Hochstrasser and Tissot 1993). All histories involved the global study of proteins in a body fluid. In several cases, this global approach led to the development of simpler diagnostic tests.

A patient with meningitis and head trauma suffered a fluid leak through one ear after a skull fracture. The fluid was identified as spinal fluid by its 2-D PAGE protein pattern and especially by the spinal fluid-specific post-translational modifications of serotransferrin. Today, this post-translational modification can be detected with simpler one-dimension gels or even a monoclonal immunoassay.

A patient with acute respiratory distress had a chest mass compressing the trachea. On computerised tomography scan, the mass was filled by fluid. After needle aspiration to relieve the patient's symptoms, 2-D PAGE was used to analyse the fluid proteins. The protein pattern corresponded to lysed old clotted blood from a previous traumatic hematoma, which ruled out any type of large abscess or cancer with central necrosis and fluid collection.

A patient with renal insufficiency was incorrectly diagnosed as having hereditary familial amyloïdosis. This amyloïdosis was in fact secondary to an IgA myeloma missed by conventional techniques because of the tremendous abundance of a severely sialylated heavy chain of a monoclonal IgA. Here, attention to the "prozone effect" of antigen excess over the reagent antibodies would have prevented this diagnostic failure.

A patient with a difficult to diagnose haemolytic anaemia was found to suffer a lymphoma which revealed itself at the beginning only as a faint monoclonal immunoglobulin light chain production.

A patient with rapidly progressing dementia was found to have spinal fluid containing two unusual proteins (numbered 130 and 131) which were recently identified as Tau γ chain, also known as protein 14–3–3. With the clinical presentation, these findings were pathognomonic of Creutzfeldt-Jakob disease (CJD). The work of Harrington and coworkers, of others (Blisard et al. 1990; Harrington and Merril 1984; Harrington and Merril 1985; Harrington and Merril 1988; Harrington et al. 1986; Hsich et al. 1996; Marzewski et al. 1988; Yun et al. 1992; Zerr et al. 1996) and this case presentation demonstrate the benefits of a proteomic approach for the discovery of new disease markers or patterns. There were no premortem biochemical diagnostic tests for CJD until the discovery that the normal brain protein Tau leaks into spinal fluid in CJD. Fig. 8.1 shows four similar regions of silver stained 2-D PAGE images. It highlights how spots 130 and 131 were identified as Tau γ chain by comparing normal brain, normal and diseased CJD spinal fluid (Hsich et al. 1996).

The proteomic approach applied here was clearly successful at identifying a new disease marker (Hsich et al. 1996). Even if tests to detect the prion protein itself in spinal fluid or in blood become available in the near future, the global measurement of normal brain proteins leaking into spinal fluid during disease progression will be of use to monitor the disease course and, hopefully, disease treatment.

8.4.1 Glycoproteins in body fluids

Modifications in the glycans on glycoproteins can occur in diseases such as cancers, inflammatory processes and alcoholism. Two-dimensional PAGE as part of a

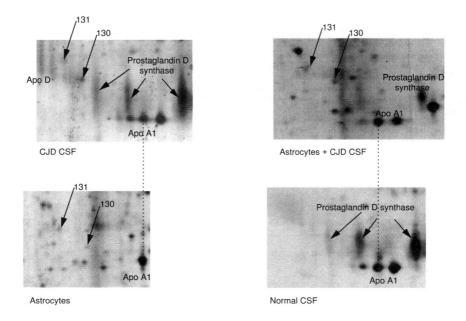

Fig. 8.1. Four similar regions of silver stained 2-D PAGE images showing how spots 130 and 131 or Tau γ chain were identified by comparing normal brain (astrocytes) to normal and diseased (CJD) cerebrospinal fluid (CSF). A mixture of diseased spinal fluid (CJD CSF) and astrocytes were analysed to verify the co-migration of the spots of interest

proteomic approach provides a refined tool to analyse the microheterogeneity of body fluids and especially plasma glycoproteins (see Chap. 4). During investigations involving plasma and serum 2-D maps, we identified several new plasma glycoproteins including the ApoJ / clusterin / SP-40,40 protein (Blatter et al. 1990; Hochstrasser et al. 1988; Hughes et al. 1992; James et al. 1988). We have also combined 2-D PAGE with lectin affinity blotting to establish which proteins carry glycans in the normal human plasma (Golaz et al. 1995; Gravel et al. 1994). Analysis of plasma/serum proteins from human foetuses and infants and their comparison with adult reference protein maps allowed tens of different proteins to be identified, including their genetic variants in some cases (Tissot et al. 1993).

In numerous diseases, the glycoprotein pattern changes in a reproducible and easily detectable manner. For example, Yuasa et al. (1995) studied the carbohydrate-deficient glycoprotein (CDG) syndrome and found increases in isoforms with higher isoelectric points in all proteins carrying N-linked glycosylation that were tested. They suggested that a deficiency of N-linked oligosaccharide transfer occurs on many serum glycoproteins in patients with CDG syndrome. Henry et al. (1997) also analysed serum glycoproteins and their liver precursors in patients with carbohydrate-deficient glycoprotein syndrome type I and IV. They found apparent deficiencies in ApoJ and serum amyloid P protein. The pattern of serum glycoproteins showed abnormal trains of isoforms with decreased mass and a cathodic shift. They postulated:

"...that these abnormal precursors accumulate during the early oligosaccharide processing of the nascent protein-bound oligosaccharides and that glycoprotein precursors undergo an altered intracellular transport while the post-translational processing along the normal pathway is still apparently functioning in patients with CDG" (Henry et al. 1997; also see Heyne et al. 1997).

Moule et al. (1987) studied the sialylation and microheterogeneity of human serum α-1 acid glycoprotein in sera of patients with rheumatoid arthritis, myocardial infarction, cancer and of healthy volunteers. They found considerable charge heterogeneity in all samples by 2-D PAGE and a higher sialic acid level in α1-acid glycoprotein purified from the sera of cancer patients.

The histories described above illustrate the proteomic approach in several body fluids in basic and applied medical research. Other biomedical applications of proteomics are in the domain of pharmaceutical research and toxicology, and in cancer diagnosis, prognosis and treatment evaluation. These are discussed in the following sections of this chapter.

8.5 Proteomics, toxicology and pharmaceuticals

The simultaneous identification, characterisation and quantitation of numerous gene products and their post-translational modifications is now mandatory to evaluate how multiple overlapping pathways are influenced by toxins or drug treatment. Several authors have already used 2-D PAGE as the core methodology for pharmaceutical and toxicological studies (Anderson 1981; Anderson and Gemmell 1984; Anderson et al. 1984; Anderson et al. 1995; Anderson et al. 1996a,b; Cunningham et al. 1995; Myers et al. 1997; Myers et al. 1995; Richardson et al. 1994; Weinstein et al. 1997; Witzmann et al. 1997), and it is anticipated that the massively parallel approach offered by proteome projects will lead to new drug discoveries. The aim of this section is to demonstrate the benefits that a proteomic approach has already provided to the domains of pharmacology and toxicology.

Quantitative protein expression changes due to drugs can be measured accurately enough to detect multigenic and overlapping pathway effects. An example of this quantitation requirement was recently published by Anderson et al. (1996a) who investigated the effects of five peroxisome proliferators on proteins in the liver of rats. They demonstrated that the tested peroxisome proliferators produced effects on protein abundance over wide time and dose ranges (Anderson et al. 1996a). To further extend these measurements of qualitative and quantitative protein expression changes, Anderson and collaborators are currently developing a Molecular Effects Database describing xenobiotic effects in rodent liver. Their database can

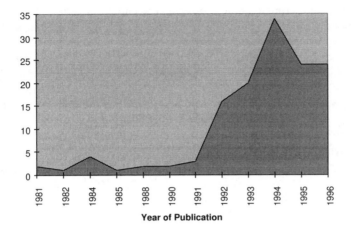

Fig. 8.2. The number of publications from 1981 to 1996 in the area of toxicology and pharmaceutical research. These numbers result from a Medline search with keywords human and 2-D PAGE and toxicology or pharmaceutical

detect, classify, and characterise a broad range of liver toxicity mechanisms. In addition, they are building companion 2-D databases describing rodent brain and kidney which have links to genomic sequence databases (Anderson et al. 1996b). As they mentioned in their publication:

"...assimilation of this approach into research and regulatory toxicology poses an interesting challenge — one that is likely to lead to a radically more sophisticated understanding of toxicity and its biological basis" (Anderson et al. 1996b).

Two-dimensional PAGE methodology has also been extensively applied to human samples with a potential clinical orientation in the pharmacological and toxicological domain. The graph in Fig. 8.2 shows the number of publications for the last 15 years in the area of toxicology and pharmaceutical research using or relating to human samples, mentioning 2-D PAGE and referenced in the Medline database. The search was restricted to human samples only, for clinical relevance and practical reasons. Clearly, there has been a recent interest in the use of 2-D PAGE as a means to undertake a combined genomic and proteomic approach. But this number is still quite small. Many more pharmacologists and toxicologists should use this approach in their work to simultaneously analyse many gene products and their post-translational modifications. Current methodology is highly reproducible and increasingly easy to perform (see Chap. 2), and several new biotechnological tools are available to rapidly characterise the separated proteins (see Chap. 3 and 4). Indeed, as many drug primary pharmacological and metabolic effects might only be post-translational in nature in several situations, a means to display many post-translational modifications simultaneously could be very useful to analyse the biochemical mechanisms involved and to approach mechanisms that induce side effects.

A small number of the publications from Fig. 8.2 are summarised below. They have been chosen to highlight the potential and need to study protein post-translational modifications within biomedical projects. We shall briefly discuss work related to retinoic acid, a fascinating drug with differentiation potential in certain leukemias. Then, work relating to the study of other covalent protein modifications induced by or related to drugs or hormones will be addressed. The discussed post-translational modifications will be glycosylation, phosphorylation and drug adducts. Particular emphasis will be placed on effects on cell growth or multiplication to introduce the latter sections of this chapter related to cancer. Finally, other pharmacological treatments and their effects will be summarised, and studies on hormones will be addressed. Related areas of cancer treatment will be discussed in Sect. 8.7.

8.5.1 Retinoic acid

Retinoic acid belongs to the retinoid family which includes vitamin A. It is used in dermatology and onco-haematology. In dermatology, the 13-cis retinoic acid formula is used to treat severe acne. Because of its important teratogenic potential, this drug is used with extreme caution and never during pregnancy. The drug is transformed *in vivo* to all trans retinoic acid.

Many of the effects of retinoic acid seem to be related to protein retinoylation, or in other words retinoic acid acylation (Takahashi and Breitman 1991; Takahashi et al. 1991; Takahashi and Breitman 1992a,b; Takahashi and Breitman 1994). Several proteins are post-translationally modified by retinoylation, such as vimentin or heat shock proteins. A link between retinoylation of proteins and glycosylation has also been suggested and some of the effects of retinoic acid might well be differences in protein fucosylation. The simultaneous measurement and characterisation of numerous proteins is required in this situation to analyse the pharmacological properties of this drug. Retinoic acid can induce cell differentiation in acute promyelocytic leukaemia. Proteome studies will be useful to establish how acute promyelocytic leukaemia responds to retinoic acid, and for how long.

In the study of a drug which has an impact on cell growth and/or differentiation, one might expect that cellular mRNA levels would parallel protein expression and pharmacological effects. Interestingly, this seems not to be the case. The complex regulation of protein expression and post-translational modifications by retinoic acid during granulocytic differentiation of human leukaemic cells has been highlighted by Spector et al. (1994), who demonstrated a lack of correlation between mRNA level and protein concentration. This lack of correlation was also demonstrated in a recent study on liver mRNA and protein levels (Anderson and Seilhamer 1997). The coefficient of correlation between mRNA levels and the respective protein abundance was only 0.48 (Fig. 8.3). This crucial experiment needs to be repeated and extended. Yet, it is likely that mRNA and respective protein abundance will never correlate completely because their respective half lives

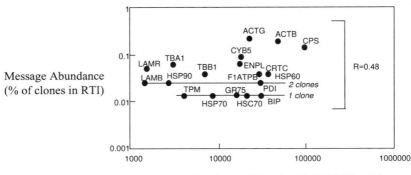

Fig. 8.3. Log-log plot of the abundance of 19 gene products. Comparison between the protein abundance (X axis) as measured by Coomassie Blue (CB) staining on 2-D PAGE gels and messenger RNA abundance (Y axis) as measured by the percentage of clones on RTI (modified from Anderson and Seilhamer 1997)

are certainly different as well as their cell trafficking. Other factors such as RNA elongation also play a role in protein expression regulation. In consequence, as stated above, a combined genomic and proteomic approach will be necessary for most pharmaceutical and toxicological studies, especially because of the crucial and extraordinarily complex post-translational modifications which occur on many proteins, such as retinoylation.

8.5.2 Glycosylation

Polypeptide glycosylation in humans is extraordinarily complex. The progressive complexity of post-translational modifications during evolution has probably developed to allow more diversity without necessarily increasing genome size (see Sect. 8.6). In human body fluids most proteins are glycosylated. Many globulins display a tremendous but tightly controlled glycosylation diversity which can change during disease processes or after drug or toxin exposure.

Two-dimensional gel electrophoresis has been central to several discoveries regarding serum glycoproteins. Lectin affinity blotting was used to establish which proteins carry glycans in the normal and diseased human plasma map. We found no evidence for protein modifications in fresh frozen plasma after photochemical treatment (Tissot et al. 1994), but in ethanol abuse, alterations of serum glycoproteins were detected including desialylation (Gravel et al. 1994; Gravel et al. 1996).

This toxic effect of ethanol is well known and leads to carbohydrate deficient glycoproteins such as carbohydrate deficient transferrin.

Not only body fluid proteins, but several hormones are also heavily glycosylated. Erythropoietin is an extreme example. Its final molecular weight is between 34 and 40 kDa while its polypeptide chain weighs only 18.4 kDa (Hammerling et al. 1996). Erythropoietin is a fascinating molecule which is synthesised by the kidney and which stimulates the production of red blood cells by the bone marrow. As mentioned in the introduction of this chapter, its recombinant analogue is used extensively in medicine today to stimulate the production of red blood cells especially in patients with end stage renal insufficiency and on dialysis. The recombinant molecule has been studied structurally using mass spectrometry (MS) and nuclear magnetic resonance spectroscopy (NMR) techniques by Rush et al. (1995). They highlighted the extraordinarily complex microheterogeneity of this molecule.

The efficacy of erythropoietin appears to depend upon its branching sugar structure. Although MS, NMR and other techniques are required to analyse the complexity of such a glycoprotein in detail, a 2-D PAGE approach combined with a biological assay provides a different means to understand the functions of this molecule. Potencies of various intact and sugar-trimmed rhEpo preparations were investigated using 2-D PAGE. Minor truncations in the rhEpo-attached glycan moieties were detected using picomolar quantities of the hormone, and the effect of carbohydrate-dependent microheterogeneity was highlighted (Hammerling et al. 1996).

Cell surface proteins are also often glycosylated. They play an important role in cell recognition and function. As an example, platelet properties and functions such as electrokinetic motion (Bateson et al. 1994) or aggregation (Hardisty et al. 1992) rely on glycoproteins, and their abnormal distribution can cause lifelong bleeding disorders. Cell-cell interactions also play a crucial role in numerous disease processes and especially in cancer invasion. Cell growth inhibition by cell-cell interaction is essential. Glycoproteins such as contactinhibin (Wieser et al. 1995) or gangliosides in melanoma, are involved in such interactions.

8.5.3 Phosphorylation

In many biochemical pathways, the "on" or "off" signals involve the phosphorylation or dephosphorylation of polypeptides by kinases and phosphatases. The pharmacological action of certain drugs or the toxic effect of certain toxins can modulate or profoundly change these phosphorylation or dephosphorylation events.

There are several studies in pharmacology and toxicology where protein phosphorylation or dephosphorylation was analysed using 2-D PAGE. For example, several cytokines interact with receptors of the hematopoietin receptor superfamily (Linnekin et al. 1992), the signal transduction pathways including several phosphorylated proteins. Bistratene A changes the protein phosphorylation patterns in human leukaemia cells (Watters et al. 1992). Many hormones trigger a cascade of phosphorylation or dephosphorylation in cells. Human chorionic gonadotropin induces protein phosphorylation in chorionic tissue (Yamawaki and Toyoda 1994). These summaries are only a few examples where the study of protein phosphorylation on human samples was done with 2-D PAGE technology.

8.5.4 Drug adducts and vehicles

Drugs or drug metabolites, like phosphate or sugar groups, can be covalently bound to proteins, therefore modifying their primary, secondary and tertiary structure. To study their effects and especially their toxicity, it is necessary to identify the modified polypeptides. A proteome approach is well suited to simultaneously identify and quantitate these new adducts.

For example, a comparative study of mouse liver proteins arylated by reactive metabolites of acetaminophen and its nonhepatotoxic regioisomer was conducted by Myers and colleagues (1995). Acetaminophen is commonly used as an analgesic and antipyretic all around the world. It is hepatotoxic and can be lethal in large doses. Its *m*-hydroxy isomer is apparently not hepatotoxic. Both are oxidised by liver cytochromes P-450 to reactive metabolites that bind covalently to hepatic proteins in a mouse and can therefore by studied by an efficient 2-D PAGE approach.

Drugs are often transported in blood bound to albumin or glycoproteins. Certain drugs get covalently bound to them. In consequence, the drug bioavailability is modified and could be dramatically decreased. For example, Marzocchi et al. (1995) demonstrated that human serum albumin is modified by reaction with penicillins.

To enhance drug delivery and improve transport, colloidal carriers or nanoparticles can be synthesised and used as drug carriers in blood. Their surfaces are engineered to deliver their contents to the selected target. Here again, a proteome approach could be useful to detect and identify the blood proteins interacting with the artificial particles over time. As an example, plasma protein adsorption patterns on surface-modified latex particles were evaluated by 2-D PAGE (Blunk et al. 1993). Several apolipoproteins seemed to play a major role in the uptake of these particles by the reticulo-endothelial system. Leroux and colleagues (1994) investigated the interactions of poly-lactic acid nanoparticles prepared by a recently developed salting-out process, with lymphocytes and monocytes isolated from healthy human donors. They tentatively identified plasma opsonins adsorbed onto the nanoparticle surface and found specific apolipoproteins as well.

8.5.5 Other pharmacological treatments

The fundamental actions of several pharmacological agents on numerous model cell lines have been studied and their effects on overall protein expression and posttranslational modifications have been demonstrated. For example, Anderson

(1981) identified mitochondrial proteins and some of their precursors in 2-D electrophoretic maps of human lymphoid cells treated with various antimitochondrial agents. Anderson and Gemmell (1984) also studied the protein pattern changes and morphological effects due to methionine starvation or treatment with 5-azacytidine of cell lines, and observed changes in the protein pattern of HL-60 promyelocytic leukaemia cells. One group of proteins was apparently modified, appearing in new positions on a 2-D gel. A further series of proteins, including several abundant nuclear polypeptides, were substantially diminished suggesting lower expression. Their results thus suggested interesting relationships between methionine metabolism, protein and structural changes in the cell nucleus, and cell-induced differentiation. Croxtall and collaborators (1988) studied protein synthesis in cancerous and normal human endometrium by the incorporation of $[^{35}S]$-methionine following medroxyprogesterone acetate treatment.

The point must be made that many, if not all of the above studies, were somewhat descriptive. However it must be appreciated that until recently there has been no practical means to characterise proteins of interest on a large scale. Today's proteome approach, which benefits from an explosion in genome data and improvements in bioinformatics, is dramatically different and it is anticipated that it will soon be possible to identify hundreds of proteins in weeks or days and to rapidly characterise protein post-translational modifications.

8.5.6 Hormones

In obstetrics and gynaecology, several hormones are currently extensively used for contraception and infertility. Several scientists have taken a proteomic approach to study the overall protein expression following hormonal treatment.

High-dose ethinyl estradiol-norgestrel emergency contraception was studied by Young et al. (1994) who analysed endometrial proteins in serum and uterine luminal fluid. They found that this treatment effectively suppressed progesterone-associated endometrial proteins in the midluteal uterus, potentially altering the endometrial environment unfavourably and affecting the survival of the early embryo.

Ho and colleagues (1994) measured the polypeptide pattern from human breast epithelial cells following human chorionic gonadotropin (hCG) treatment. They stated that at least 11 proteins were preferentially synthesised and five specific polypeptides were decreased in hCG treated cells in comparison with controls. With today's technology, these 16 polypeptides could probably be substantially microsequenced by mass spectrometry directly from gels. In their experiment, the hCG induced at least four new mRNAs which encoded protein in the molecular mass range of 24–72 kDa. It also increased the expression of at least six mRNAs and reduced the expression of least four mRNAs in comparison with control cells. The hCG-treated cells actively synthesised a 33 kDa polypeptide which was not present in control cells.

8.6 Generation of complexity in proteomes

Chapter 4 and the publications discussed in Sect. 8.5 highlight the extraordinary complexity and essential nature of protein post-translational modifications. In a sense, protein post-translational modifications are a means of increasing proteome complexity without a corresponding increase in genome size. To display many post-translational modifications simultaneously in a graphical manner, we have defined an approach which we have called "vector maps" (Wilkins et al. 1996b). In denaturing conditions in immobilised pH gradients, the migration position of polypeptides which are not post-translationally modified can be predicted with high accuracy (Bjellqvist et al. 1993; Bjellqvist et al. 1994). As such, it is possible to generate a theoretical protein 2-D map and to compare it with the experimental separation. Vectors produced by image matching on a computer program such as Melanie II can highlight protein post-translational modifications.

Theoretical gene product maps with a hundred or so genes in *Escherichia coli*, *Saccharomyces cerevisiae*, human body fluids and human cell lines were linked to their real protein spots on 2-D PAGE (Fig. 8.4, see p. 147). On average three genes in *E. coli* produced four 2-D PAGE spots, one gene of *S. cerevisiae* three spots, one human gene between three and more than ten spots for the secreted proteins (Wilkins et al. 1996b). Despite the relatively low number of genes and respective proteins studied, it is becoming clear that the majority of cellular and secreted human proteins are probably post-translationally modified. A differential post-translational vector map approach could be powerful to study diseases, their treatment and potential side effects of drugs.

8.7 Proteomics and cancer

8.7.1 Cell cycle and cancer theory

Carcinogenic products act similarly to pharmaceutical agents, affecting the post-translational modifications and the level of expression of numerous proteins. This section will summarise some aspects of proteomics and cancer studies.

Numerous chemicals can induce cancer. But multiple genetic modifications or changes are required for cancer development. Many of the drugs lead to genetic instability and also to altered DNA repair mechanisms. Recently the mutation of genes which encode for components of cell cycle checkpoints were shown to increase genetic instability and induce cancer processes. Not only oncogene and oncogene product alterations, but cell cycle specific protein modifications play a major role in tumorogenesis and cancer progression (Hartwell and Kastan 1994). Completion of the cell cycle depends upon the coordination of numerous macro-molecular synthetic assemblies and migratory processes. A complex procedure involving changes in kinase and cyclin components drives the cell from one step to the next in the cell cycle. Transcription of cyclin genes, degradation of cyclin proteins and the phosphorylation or post-translational modification of several enzymes control the passage of the cells through the cycle and act as control checkpoints. Only a combined genomic and proteomic approach will show the key proteins and their complex regulations and interactions.

Both positive and negative controls guide the cell cycle progression (Nasmyth 1993). In the case of genome or DNA damage, the cell cycle and cell proliferation can be stopped by negative control loops. The damaged cell then undergoes apoptosis or programmed cell death. Compromised negative feedback would allow proliferation of abnormal cells with increasing DNA instability and damage.

It is quite obvious that cell cycle dysfunction or dysregulation and cancer in humans result from complex gene interactions, numerous cellular events and environmental influences. In addition, every patient is genetically and thus phenotypically unique. Therefore, the precise diagnosis and prognosis in a proliferative disorder will depend on the study of epigenetic regulation of tissue survival and growth for each patient. It will require a powerful and efficient proteome approach.

The above summary underscores the complexity of cancer development. The following paragraphs will be less theoretical, focusing more on how a proteomic approach can be used to study cancerous processes within a clinical environment. The contributions that such an approach has already made to cancer studies will be shown. Five areas will be addressed, namely i) sample preparation, ii) tissue of origin, iii) the metastatic potential and cancer treatments, iv) data analysis and automatic diagnosis and v) cell cycle proteins and oncogenes.

8.7.2 Sample preparation

The reliable measurement of protein expression in patient cells or tissue biopsies relies almost entirely on the quality of sample preparation (see Chap. 2). Franzen and collaborators have addressed this critical step in several publications (Franzen et al. 1991; Franzen et al. 1993; Franzen et al. 1995). They demonstrated that nonenzymatic methods for the preparation of tumour cells, including fine needle aspiration, scraping or squeezing tissue biopsies, had advantages over methods using enzymatic extraction of cells. Nonenzymatic methods were shown to be rapid and to reduce loss of high molecular weight proteins. These methods did not require the separation of viable and nonviable cells by Percoll gradient centrifugation. They also analysed qualitative aspects of tissue preparation in relation to the histopathology of lung cancer, and examined the relationship between histopathological findings and 2-D PAGE gel quality. They concluded that histopathological features, such as a local homogeneity, the amounts of connective tissue and serum proteins were critical factors for the successful preparation of the sample and the high quality of overall protein separation and analysis. They clearly overcame some major technical difficulties. As a result of their work, clear guidelines are now available for sample preparation of patient cells and biopsies (Franzen et al. 1995).

Similarly, Reymond et al. (1997) designed and compared specific sample preparations in colorectal cancer. The epithelial cell content in colorectal cancer and normal mucosa showed important intersample variations. A method allowing the preparation of pure epithelial cell samples from normal and tumoral colonic fresh mucosa was proposed.

8.7.3 Tissue of origin

From a clinical perspective, the identification of the tissue of origin is often mandatory to define a diagnosis as well as a prognosis and choice of treatment. Even with currently available immunochemistry methods, pathologists sometimes have difficulties to identify a tissue under a microscope or the tissue of origin of, for example, an adenocarcinoma; and hence the designation adenocarcinoma of unknown primary or unknown origin. The following example highlights the potential of a proteome approach to overcome this difficulty. A patient was found to have an abdominal tumour. The mass was surgically resected. But the physicians had difficulties distinguishing between duodenal and pancreatic cancer. Therefore, samples of the tumour, normal duodenal mucosa and normal pancreatic tissue from the same patient were compared by 2-D PAGE. The tumour pattern resembled the normal duodenal mucosa more closely than pancreatic tissue, suggesting that the tumour was of duodenal origin (Isoda et al. 1990).

Perhaps one of the most suitable medical applications of a proteome approach is in the study of cancerous tissues. In many cases it is possible to make a direct comparison of proteins expressed in normal and diseased tissue from a single patient. The comparison of 2-D separations of such tissues can immediately highlight proteins that are present in greater or lesser quantities, new proteins expressed only in the cancerous cells, and changes in protein post-translational modifications. Below are reviewed studies on cancerous tissues using a proteome approach. They are reviewed in an anatomical order from head to toes and with emphasis on data analysis and management. The tissues discussed are brain, thyroid, breast, lung, colon, kidney, bladder, ovary and bone marrow.

Brain. Hanash et al. (1985) studied protein patterns of nine neuroblastoma and thirteen other cell lines. They scored more than 600 polypeptides and found one which was present in all neuroblastoma cell lines but none of the other cell types. In 1996 they extended their study with an elegant combined genomic and proteomic approach using 2-D separations of neuroblastoma DNA and proteins. The benefit of their two 2-D approaches was, as they said:

"...the efficiency of scale and the ease with which abundant proteins or multicopy genomic fragments could be detected, identified and quantitatively analysed" (Wimmer et al. 1996).

Thyroid. Lin et al. (1995) compared membrane proteins from benign and malignant human thyroid tissues. Their goal was to improve the differential diagnosis of thyroid malignancy.

Breast. Specific protein differences in normal versus malignant human breast tissues have been detected by Wirth and collaborators (1987). They found that at least 22 polypeptides were significantly and consistently increased in tumour samples while only one polypeptide was decreased. They also found a relationship between the expression of oestrogen and progesterone receptors and certain polypeptides.

Giometti et al. (1997) have extensively studied breast cancerous cell lines and are currently developing a database of human breast epithelial cell proteins using quantitative 2-D PAGE. They determined which proteins are present in different types of human breast cells (milk producing and nonproducing, oestrogen receptor positive and negative, normal and malignant) and which proteins change in abundance in response to stimuli that trigger cell differentiation, growth, or death. Rasmussen et al. (1997) have also analysed human breast carcinoma proteins and studied proteins that bind to specific ligands.

Lung. In many countries, lung cancer is the number one oncological disease. Often the diagnosis is so late that no curative measure exists. Chemotherapy regimens are often barely palliative for these aggressive solid tumours. Therefore, any progress in the diagnostic or prognostic evaluation and in the modes of therapy would be very beneficial. Several authors have studied human lung cancer using 2-D PAGE technology.

Fourteen lung tumours of various histopathological types were analysed by Okuzawa et al. (1994). They found a variation in the expression of several proteins which correlated with different histological types. They also found a protein which was significantly overexpressed in primary lung adenocarcinomas compared to small cell lung carcinomas, squamous cell lung carcinomas, metastatic lung adenocarcinomas from colon and rectum, and normal tissue. They concluded that their approach was effective to define new tumour-specific markers. This work is significant as some clinicians believe that the current distinctions between squamous cell carcinoma and adenocarcinoma are of no real value and thus new predictive and prognostic markers would be useful.

Hirano et al. (1995) analysed the relationship between the histopathological findings in primary lung malignancies and the expression of a number of unidentified polypeptides. Forty-five samples from patients with primary lung cancer, including 21 adenocarcinoma, ten squamous cell carcinoma, five large-cell carcinoma, one adenosquamous cell carcinoma, five small-cell carcinoma and three carcinoid tumour were examined. Sixteen polypeptides were thought to be associated with histopathological features. The authors thought that it was possible to classify primary lung carcinomas based on the 2-D PAGE findings more precisely than on morphology alone (see also Sect. 8.7.5).

Colon. Colon cancers are also very frequent. Despite Duke's classification and a properly designed staging technique, it is often difficult to evaluate the extent of disease at the time of diagnosis and to establish an adequate prognosis. It is difficult to know which patient has a malignant disease that has already spread to lymph nodes, liver and/or lung. Proteome studies are therefore critical for this very common disease of the elderly.

To search for new colon cancer antigens and markers, Ward et al. (1990) isolated a number of proteins from a crude cell extract of a human colon carcinoma cell line by 2-D PAGE and blotting techniques. Selected protein spots of interest were excised and subjected to Edman degradation (see also Ji et al. 1997a,b). Kovarova and colleagues (1994), in collaboration with our group, studied normal colon mucosa and colorectal cancer. Protein spots could be classified into two groups: common for normal colon mucosa and tumour, and specific for tumour tissue.

Stulik et al. (1997) analysed the different expression of colon heat shock proteins in human colonic diseases using monoclonal antibodies. Immunostaining of hsp70 in polypous and malignant tissues showed qualitative and quantitative changes in the expression of the more acidic isoforms. Also, the different basic isoforms of hsp70 were detected in chronically inflamed colonic mucosa such as ulcerative colitis or Crohn's disease. Additional proteins were variably immunostained in normal and pathological specimens.

Ji and coworkers are currently establishing a 2-D gel database of human colon carcinoma cells (Ji et al. 1997a,b). Their database currently lists cellular proteins from normal crypts and from several colorectal cancer cell lines.

Kidney. Several types of kidney tumours have been described. They are all quite rare. The human kidney cancer, arising from the proximal tubular epithelium and called renal cell carcinoma, accounts for about 3% of adult malignancies. Unfortunately, as for many solid tumours, it is almost always discovered late when surgery is no longer curative. Molecular and cytogenetic analysis have highlighted deletions, translocations, or loss of heterozygosity. No marker for early renal cell carcinoma diagnosis is available today despite several attempts to find phenotypic markers.

Accordingly, we have established a human kidney protein map in the SWISS-2DPAGE database (Sarto et al. 1997). Of the 2,789 separated polypeptides, 43 were identified by gel comparison, amino acid analysis, N-terminal sequencing, and/or immunodetection. Four polypeptides were found to be absent in renal cell carcinoma and present in normal kidney tissue. One of them was identified as ubiquinol cytochrome c reductase. A second polypeptide was identified as mitochondrial NADH-ubiquinone oxido-reductase complex I. These results and others indicate that mitochondrial dysfunction might play a major role in renal cell carcinoma genesis or evolution. In this study, no specific marker proteins have yet been identified, but there are several candidate proteins that now await further characterisation.

Bladder. Bladder cancer is a dreadful disease because of its highly recurrent behaviour and multifocal origin. However, it has one potential advantage: one might expect to find potential markers in the urine.

For this reason, Rasmussen and coworkers (1996) are establishing a comprehensive 2-D database of proteins from the urine of patients with bladder cancer. Their urine protein database includes all the polypeptides detected in the urine of 50 patients and contains 339 proteins of which 124 have been already identified. Several other proteins of interest have been discovered but remain unknown in databases. It is hoped that markers may be identified which can serve as prognostic factors (Rasmussen et al. 1996).

In this direction, Celis et al. (1996a) have discovered a loss of adipocyte-type fatty acid binding protein and other protein biomarkers in association with progression of human bladder transitional cell carcinomas. In addition, they suggested that adipocyte-type fatty acid binding protein is an important component of the pathway(s) leading to bladder cancer development. In another study, four patients with squamous cell carcinoma of the bladder were identified from 100 samples of patients with suspected transitional cell carcinoma by measuring adipocyte-type fatty acid binding protein in the urine (Celis et al. 1996b).

Ovary. Several types of ovarian cancer exist. The more malignant tumours often spread to the abdominal cavity and at a later stage are responsible for malignant ascitic fluid accumulation. Nucleic-acid and protein-based approaches should be undertaken to better characterise these diseases.

Quantitative protein changes were measured and compared in metastatic and primary epithelial ovarian carcinoma by 2-D techniques (Lawson et al. 1991). They found that the abundance of two proteins was significantly and constantly decreased in metastatic tumour cells in comparison to primary tumour cells.

Bone marrow. Hanash and coworkers have extensively studied childhood leukaemia for a number of years (Hanash and Baier 1986; Hanash et al. 1982; Hanash et al. 1986a,b; Hanash et al. 1988; Hanash et al. 1989; Hanash et al. 1993). Two-dimensional PAGE allowed them to detect 11 polypeptide markers that distinguished between subtypes of acute lymphoblastic leukaemia and between acute lymphoblastic and acute myelocytic leukaemia. In two children with otherwise undifferentiated leukaemia, they found cellular proteins revealing myeloid origin of the blast cells. They also found a new marker for common acute lymphoblastic leukaemia.

Keim et al. (1990) studied the proliferating cell nuclear antigen expression in childhood acute leukaemia and also found differences in its expression according to leukaemia subtype. They believed that these differences were not related to the initial peripheral white blood count, age, or sex, but that they reflected differences in proliferative activity between subtypes of acute leukaemia.

Saunders et al. (1993) studied the protein expression in chronic lymphocytic leukaemia cells and normal B-lymphocytes. Protein synthesis was analysed in leukaemic cells from 10 chronic lymphocytic leukemias after [^{14}C]-radiolabelling. They found only minor differences between each of the chronic lymphocytic leukaemia samples. The expression level of some proteins might be correlated to the stage of the disease.

8.7.4 Metastatic potential and cancer treatment

The evaluation of the metastatic potential of tumours remains difficult and it is critical that results, which are used to determine the selection and aggressiveness of the treatment, are always precise and of high quality. A study in this direction was done by Schwalke et al. (1990). Using a 2-D PAGE approach, they studied the protein differences in human pancreatic cancer cell lines with diverse metastatic potential. Similarly, Osada et al. (1996) studied the role of E-cadherin in the intrahepatic metastasis of hepatocellular carcinoma.

For the evaluation of treatment, Weinstein et al. (1997) have established an information-intensive approach to the molecular pharmacology of cancer. Since 1990, the American National Cancer Institute has screened more than 60,000 compounds against a panel of 60 human cancer cell lines. Chemical compounds and a number of natural product extracts have been examined for their ability to inhibit the growth of cancer cell lines representing different organs of origin. Myers et al. (1997) are using the above screening program to look at the protein expression of the cells in response to the various compounds. They have developed a 2-D PAGE protein expression database covering all 60 cell lines, and have already analysed the correlation among protein spots in terms of their patterns of expression and in terms of their apparent relationships to the pharmacology of a set of 3,989 screened compounds.

8.7.5 Data analysis and automatic diagnosis

There is no doubt that the interpretation of 2-D PAGE images can be facilitated by statistical methods, artificial intelligence and machine learning programs. The correspondence analysis approach and ascendant hierarchical classification differ significantly from the more classical approach of principal component analysis. Starting with a series of gels, each having a large number of spots, correspondence analysis provides for the representation of the samples in a factorial space of reduced dimension. Ascendant hierarchical classification sorts the images into meaningful groups. Simultaneous representation of both polypeptide spots and gel images are done in the same factorial space (Pun et al. 1988). The characteristic proteins representative of a particular class of gels (e.g. lung cancer samples) are precisely highlighted, greatly simplifying the analysis of gel images.

A multivariate statistical approach was applied to classify lung tumour cells via their 2-D protein pattern by Schmid et al. (1995). They used correspondence analy-

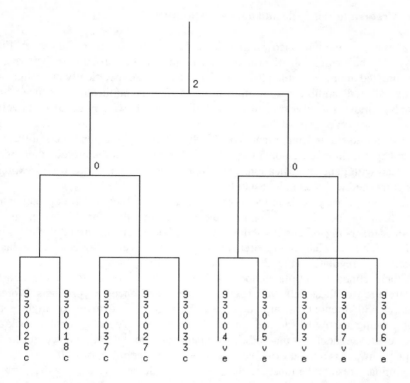

Fig. 8.6. Descending hierarchical tree. Heuristic clustering of young (c) and aging (ve) rat endothelial cell 2-D PAGE silver stained images in an atherosclerosis study by Cremona et al. (1995)

sis and ascendant hierarchical clustering. Their results indicated that protein patterns are highly typical of cell types. It also allowed the identification of new potential marker proteins.

The ultimate goal of our Melanie project in Geneva is to go directly from a 2-D separation of a tissue biopsy to automatic protein map interpretation by computer (Appel et al. 1991). Accordingly, the software we have developed can automatically classify protein patterns using principle component analysis (Fig. 8.5, see p. 148), wavelet decomposition, neural networks and heuristic clustering analysis (Fig. 8.6), the first and last being so far incorporated into our Melanie software program (Appel et al. 1988; Wilkins et al. 1996a; Zahnd et al. 1994; Zahnd et al. 1993).

8.7.6 A return to cell cycle and oncogene products

Whilst the above mentioned studies have generally examined cancer to find changes in the expression of proteins and define marker proteins or patterns, a more directed approach could also be undertaken to look specifically at changes in cyclins, cdk, cdk inhibitors and other proteins already implicated in cancer. For example, almost fifteen years ago Celis et al. (1984) showed evidence that cyclin was a central component of the pathway(s) that controls cell proliferation. More recently, in one study, Conner and Wirth (1996) analysed the genomic stability of human hepatocellular carcinoma HepG2 cells and measured the protein alterations associated with gene amplification. They found seven polypeptides whose expressions correlated with genomic instability.

One of our primary goals in the proteomic study of cancer is to find protein patterns which are specific for malignant cancer behaviour, therefore having prognostic value and being potentially informative for treatment selection. With this in mind, we have established a theoretical protein map of several oncogene products and cell cycle regulators (Fig. 8.7; Sanchez et al. 1997).

We then refined a multiple monoclonal immunoblot technique to simultaneously measure a few of these cell cycle specific proteins and oncogene products. Protein detection with monoclonal antibodies are the protein equivalent to single gene studies, being narrowly focused on single molecules. An immunoblot cocktail using several monoclonal antibodies simultaneously to determine a series of signalling pathways is a good example of a powerful proteomic approach.

Preliminary results demonstrate that we can simultaneously display the expression and post-translational modifications of these gene products (Sanchez et al. 1997). As shown in Fig. 8.8, these analyses detected several low abundance proteins and demonstrated their extensive expression changes and post-translational modifications. It is very unlikely that these modifications were artefacts due to the separation technique. As about 30% of cellular proteins are likely to be phosphorylated, a large number of these modifications are thought to be due to the addition of phosphate groups.

Our motivation to measure numerous gene products simultaneously in cancer is not only to find new disease markers but is driven by the following statement:

"We can anticipate a time when it will be possible to characterise tumours individually for their checkpoint and repair status and thereby predict their response to particular therapies" (Hartwell and Kastan 1994).

Indeed, we are in great need of better tests in laboratory medicine to help identify patients who require selected and targeted therapy.

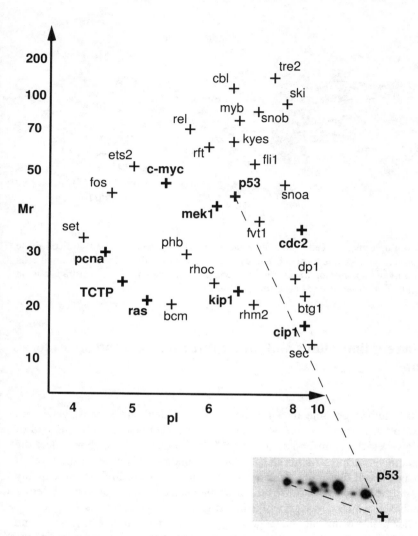

Fig. 8.7. Theoretical 2-D PAGE map modified from Sanchez et al. (1997), showing the positions of some oncogene products and cell cycle specific proteins. A small vector map (bottom right) displays the theoretical position of p53 and the links with the 10 main isoforms detected by immunoblot and chemiluminescence

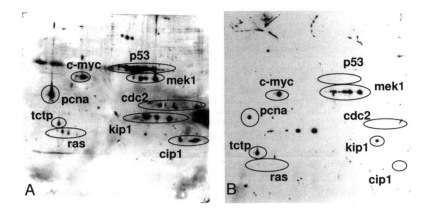

Fig. 8.8. Multiple monoclonal immunoblot detection of oncogene products and cell cycle specific proteins in (A) transformed hepatocytes and (B) normal liver. The transformed cells show not only overexpression of several oncogene products but numerous post-translational modifications. Modified from Sanchez et al. (1997)

8.8 Current limitations and future directions of proteomics for medicine

Whilst the human genome may not be sequenced until 2005, most genes are expected to be present in expressed sequence tag (EST) databases by the end of 1997. However, there are limitations to EST databases insofar as they reflect the abundance level of mRNAs in the cell. As mentioned in this chapter, the level of mRNAs in the cell does not correlate well with the abundance of proteins. In any case, it will be difficult to know if the low abundance proteins are represented in EST databases, which may hamper the investigations of such proteins.

A further major technical challenge that remains is how to detect, identify and measure simultaneously on a large scale, reproducibly and at a low cost, the expression patterns of many genes and the post-translational modifications of their products. In many situations, 2-D gels currently display probably only the "tip of the iceberg" with respect to the entire protein complement of a tissue or body fluid. If one estimates the number of human genes to be around 100,000 and one gene expresses one protein, then 100,000 gene products or proteins might be found in the human body. Now, let us consider for a moment the blood or plasma. Almost any cell type in the body which dies releases its contents into the blood. Thus in principle it is possible that most human proteins are found in the blood at some time. What about tissues or cells? Many believe that only 10% of the genes from the genome are expressed in one cell type. Then it is likely that each cell contains about 10,000 different gene products or proteins.

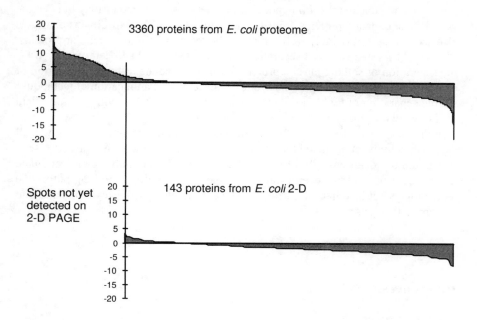

Fig. 8.9. Grand average hydropathy scores calculated according to Kyte and Doolittle (1982) for proteins from *E. coli*. The graph y axis indicates if a protein is hydrophobic (positive score) or hydrophilic (negative score). Top: Hydropathy values of all genomic *E. coli* proteins in SWISS-PROT release 34, defining the range of hydropathy for proteins from this species. Bottom: Hydropathy values for spots from *E. coli* identified on 2-D PAGE maps. Standard 2-D separation techniques as used here will separate about 85% of proteins from cells (data from Wilkins et al. unpublished)

The real situation is even more complex. Our preliminary vector maps show on average 1.3 proteins per gene in *E. coli*, 3 proteins per gene in *S. cerevisiae* and perhaps more than 10 proteins per gene in humans if one also considers body fluids. If 10,000 genes are expressed in a cell type and on average each gene product has 3 possible post-translational modifications such as 2 phosphates, one phosphate or none, then a cell might contain a total of about 30,000 different polypeptides.

Now, routine 20 by 20 cm 2-D gels can separate around 3,000 polypeptides. If one designs gels of 1 pH unit for the first dimensional separation and if each of these gels can separate 3,000 polypeptides, then the combination of 9 preparative gels from pH 3 to pH 12, should separate 27,000 polypeptides. We are not far from 30,000!

Two-dimensional PAGE offers sufficient separating power to visualise almost all proteins from a cell or tissue. But we must also consider protein abundance and hydropathy. The difference between the most and least abundant protein in a cell is certainly much smaller than in body fluid, being perhaps about five orders of magnitude. We are excited that immunodetection methods allow the display of even the very low abundance proteins from human cells (Fig. 8.8). Two-dimensional gels also seem to be surprisingly well suited for the separation of most cellular proteins. Fig. 8.9 shows that standard separation techniques, which had been optimised for the separation of soluble plasma proteins, will prove suitable for the separation of approximately 85% of the proteins from whole cells. The application of techniques for the separation of highly insoluble proteins (see Sect. 2.2.1) will certainly yield useful results for the remaining hydrophobic proteins.

In this chapter we have seen that elements of proteome technology are now well established in the clinical and biomedical research environment. We have also seen that a number of diseases have been subjected to 2-D gel analysis and candidate diagnostic proteins have been localised on the gels. It will not be long before the technologies outlined in Chap. 2 to 7 will be used to identify the key molecules and their modifications.

References

Anderson L (1981) Identification of mitochondrial proteins and some of their precursors in two-dimensional electrophoretic maps of human cells. Proc Natl Acad Sci USA 78:2407–2411

Anderson L, Seilhamer J (1997) A comparison of selected mRNA and protein abundances in human liver. Electrophoresis 18:533–537

Anderson NL, Esquer-Blasco R, Hofmann JP, Meheus L, Raymackers J, Steiner S, Witzmann F, Anderson NG (1995) An updated two-dimensional gel database of rat liver proteins useful in gene regulation and drug effect studies. Electrophoresis 16:1977–1981

Anderson NL, Esquer-Blasco R, Richardson F, Foxworthy P, Eacho P (1996a) The effects of peroxisome proliferators on protein abundances in mouse liver. Toxicol Appl Pharmacol 137:75–89

Anderson NL, Gemmell MA (1984) Protein-pattern changes and morphological effects due to methionine starvation or treatment with 5-azacytidine of the phorbol-ester-sensitive cell lines HL-60, CCL-119, and U-937. Clin Chem 30:1956–1964

Anderson NL, Hofmann JP, Gemmell A, Taylor J (1984) Global approaches to quantitative analysis of gene-expression patterns observed by use of two-dimensional gel electrophoresis. Clin Chem 30:2031–2036

Anderson NL, Taylor J, Hofmann JP, Esquer-Blasco R, Swift S, Anderson NG (1996b) Simultaneous measurement of hundreds of liver proteins: application in assessment of liver function. Toxicol Pathol 24:72–76

Appel R, Hochstrasser D, Roch C, Funk M, Muller AF, Pellegrini C (1988) Automatic classification of two-dimensional gel electrophoresis pictures by heuristic clustering analysis: a step toward machine learning. Electrophoresis 9:136–142

Appel RD, Hochstrasser DF, Funk M, Vargas JR, Pellegrini C, Muller AF, Scherrer JR (1991) The MELANIE project: from a biopsy to automatic protein map interpretation by computer. Electrophoresis 12:722–735

Bateson EA, Crook MA, Brozovic B, Crawford N (1994) Electrokinetic properties of human cryopreserved platelets. Transfus Med 4:213–219

Bjellqvist B, Hughes GJ, Pasquali C, Paquet N, Ravier F, Sanchez JC, Frutiger S, Hochstrasser D (1993) The focusing positions of polypeptides in immobilized pH gradients can be predicted from their amino acid sequences. Electrophoresis 14:1023–1031

Bjellqvist B, Basse B, Olsen E, Celis JE (1994) Reference points for comparisons of two-dimensional maps of proteins from different human cell types defined in a pH scale where isoelectric points correlate with polypeptide compositions. Electrophoresis 15:529–539

Blatter MC, James RW, Borghini I, Martin BM, Hochstrasser AC, Pometta D (1990) A novel high-density lipoprotein particle and associated protein in rat plasma. Biochim Biophys Acta 1042:19–27

Blisard KS, Davis LE, Harrington MG, Lovell JK, Kornfeld M, Berger ML (1990) Pre-mortem diagnosis of Creutzfeldt-Jakob disease by detection of abnormal cerebrospinal fluid proteins. J Neurol Sci 99:75–81

Blunk T, Hochstrasser DF, Sanchez JC, Muller BW, Muller RH (1993) Colloidal carriers for intravenous drug targeting: plasma protein adsorption patterns on surface-modified latex particles evaluated by two-dimensional polyacrylamide gel electrophoresis. Electrophoresis 14:1382–1387

Brenner S, Dove W, Herskowitz I, Thomas R (1990) Genes and development: molecular and logical themes. Genetics 126:479–486

Celis JE, Fey SJ, Larsen PM, Celis A (1984) Expression of the transformation-sensitive protein "cyclin" in normal human epidermal basal cells and simian virus 40-transformed keratinocytes. Proc Natl Acad Sci USA 81:3128–3132

Celis JE, Ostergaard M, Basse B, Celis A, Lauridsen JB, Ratz GP, Andersen I, Hein B, Wolf H, Orntoft TF, Rasmussen HH (1996a) Loss of adipocyte-type fatty acid binding protein and other protein biomarkers is associated with progression of human bladder transitional cell carcinomas. Cancer Res 56:4782–4790

Celis JE, Rasmussen HH, Vorum H, Madsen P, Honoré B, Wolf H, Orntoft TF (1996b) Bladder squamous cell carcinomas express psoriasin and externalize it to the urine. J Urol 155:2105–2112

Conner EA, Wirth PJ (1996) Protein alterations associated with gene amplification in cultured human and rodent cells. Electrophoresis 17:1257–1264

Cremona O, Muda M, Appel RD, Frutiger S, Hughes GJ, Hochstrasser DF, Geinoz A, Gabbiani G (1995) Differential protein expression in aortic smooth muscle cells cultured from newborn and aged rats. Exp Cell Res 217:280–287

Croxtall JD, Elder MG, White JO (1988) Progestin regulation of protein synthesis in endometrial cancer. J Steroid Biochem 31:207–211

Cunningham ML, Pippin LL, Anderson NL, Wenk ML (1995) The hepatocarcinogen methapyrilene but not the analog pyrilamine induces sustained hepatocellular replication and protein alterations in F344 rats in a 13-week feed study. Toxicol Appl Pharmacol 131:216–223

Franzen B, Hirano T, Okuzawa K, Uryu K, Alaiya AA, Linder S, Auer G (1995) Sample preparation of human tumors prior to two-dimensional electrophoresis of proteins. Electrophoresis 16:1087–1089

Franzen B, Iwabuchi H, Kato H, Lindholm J, Auer G (1991) Two-dimensional polyacrylamide gel electrophoresis of human lung cancer: qualitative aspects of tissue preparation in relation to histopathology. Electrophoresis 12:509–515

Franzen B, Linder S, Okuzawa K, Kato H, Auer G (1993) Nonenzymatic extraction of cells from clinical tumor material for analysis of gene expression by two-dimensional polyacrylamide gel electrophoresis. Electrophoresis 14:1045–1053

Giometti CS, Anderson NG (1984) Protein changes in activated human platelets. Clin Chem 30:2078–2083

Giometti CS, Williams K, Tollaksen SL (1997) A two-dimensional electrophoresis database of human breast epithelial cell proteins. Electrophoresis 18:573–581

Golaz O, Gravel P, Walzer C, Turler H, Balant L, Hochstrasser DF (1995) Rapid detection of the main human plasma glycoproteins by two-dimensional polyacrylamide gel electrophoresis lectin affinoblotting. Electrophoresis 16:1187–1189

Gravel P, Golaz O, Walzer C, Hochstrasser DF, Turler H, Balant LP (1994) Analysis of glycoproteins separated by two-dimensional gel electrophoresis using lectin blotting revealed by chemiluminescence. Anal Biochem 221:66–71

Gravel P, Walzer C, Aubry C, Balant LP, Yersin B, Hochstrasser DF, Guimon J (1996) New alterations of serum glycoproteins in alcoholic and cirrhotic patients revealed by high resolution two-dimensional gel electrophoresis. Biochem Biophys Res Commun 220:78–85

Hammerling U, Kroon R, Wilhelmsen T, Sjodin L (1996) In vitro bioassay for human erythropoietin based on proliferative stimulation of an erythroid cell line and analysis of carbohydrate-dependent microheterogeneity. J Pharm Biomed Anal 14:1455–1469

Hanash SM, Baier LJ (1986) Two-dimensional gel electrophoresis of cellular proteins reveals myeloid origin of blasts in two children with otherwise undifferentiated leukemia. Cancer 57:1539–1543

Hanash SM, Baier LJ, McCurry L, Schwartz SA (1986a) Lineage-related polypeptide markers in acute lymphoblastic leukemia detected by two-dimensional gel electrophoresis. Proc Natl Acad Sci USA 83:807–811

Hanash SM, Baier LJ, Welch D, Kuick R, Galteau M (1986b) Genetic variants detected among 106 lymphocyte polypeptides observed in two-dimensional gels. Am J Hum Genet 39:317–328

Hanash SM, Gagnon M, Seeger RC, Baier L (1985) Analysis of neuroblastoma cell proteins using two-dimensional electrophoresis. Prog Clin Biol Res 175:261–268

Hanash SM, Kuick R, Nichols D, Stoolman L (1988) Quantitative analysis of a new marker for common acute lymphoblastic leukemia detected by two-dimensional electrophoresis. Dis Markers 6:209–220

Hanash SM, Kuick R, Strahler J, Richardson B, Reaman G, Stoolman L, Hanson C, Nichols D, Tueche HJ (1989) Identification of a cellular polypeptide that distinguishes between acute lymphoblastic leukemia in infants and in older children. Blood 73:527–532

Hanash SM, Strahler JR, Chan Y, Kuick R, Teichroew D, Neel JV, Hailat N, Keim DR, Gratiot-Deans J, Ungar D, Melham R, Zhu XX, Andrews P, Lottspeich F, Eckerskorn C, Chu E, Ali I, Fox DA, Richardson BL, Turka LA (1993) Data base analysis of protein expression patterns during T-cell ontogeny and activation. Proc Natl Acad Sci USA 90:3314–3318

Hanash SM, Tubergen DG, Heyn RM, Neel JV, Sandy L, Stevens GS, Rosenblum BB, Krzesicki RF (1982) Two-dimensional gel electrophoresis of cell proteins in childhood leukemia, with silver staining: a preliminary report. Clin Chem 28:1026–1030

Hardisty R, Pidard D, Cox A, Nokes T, Legrand C, Bouillot C, Pannocchia A, Heilmann E, Hourdille P, Bellucci S, Nurden A (1992) A defect of platelet aggregation associated with an abnormal distribution of glycoprotein IIb-IIIa complexes within the platelet: the cause of a lifelong bleeding disorder. Blood 80:696–708

Harrington MG, Merril CR (1984) Two-dimensional electrophoresis and "ultrasensitive" silver staining of cerebrospinal fluid proteins in neurological diseases. Clin Chem 30:1933–1937

Harrington MG, Merril CR (1985) Additional cerebrospinal fluid proteins found in schizophrenia and Creutzfeldt-Jakob disease. Psychopharmacol Bull 21:361–364

Harrington MG, Merril CR (1988) Cerebrospinal fluid protein analysis in diseases of the nervous system. J Chromatogr 429:345–358

Harrington MG, Merril CR, Asher DM, Gajdusek DC (1986) Abnormal proteins in the cerebrospinal fluid of patients with Creutzfeldt-Jakob disease. N Engl J Med 315:279–283

Hartwell LH, Kastan MB (1994) Cell cycle control and cancer. Science 266:1821–1828

Henry H, Tissot JD, Messerli B, Markert M, Muntau A, Skladal D, Sperl W, Jaeken J, Weidinger S, Heyne K, Bachmann C (1997) Microheterogeneity of serum glycoproteins and their liver precursors in patients with carbohydrate-deficient glycoprotein syndrome type I: apparent deficiencies in clusterin and serum amyloid P. J Lab Clin Med 129:412–421

Heyne K, Henry H, Messerli B, Bachmann C, Stephani U, Tissot JD, Weidinger S (1997) Apolipoprotein J deficiency in types I and IV carbohydrate-deficient glycoprotein syndrome (glycanosis CDG). Eur J Pediatr 156:247–248

Hirano T, Franzen B, Uryu K, Okuzawa K, Alaiya AA, Vanky F, Rodrigues L, Ebihara Y, Kato H, Auer G (1995) Detection of polypeptides associated with the histopathological differentiation of primary lung carcinoma. Br J Cancer 72:840–848

Ho TY, Russo J, Russo IH (1994) Polypeptide pattern of human breast epithelial cells following human chorionic gonadotropin (hCG) treatment. Electrophoresis 15:746–750

Hochstrasser AC, James RW, Martin BM, Harrington M, Hochstrasser D, Pometta D, Merril CR (1988) HDL particle associated proteins in plasma and cerebrospinal fluid: identification and partial sequencing. Appl Theor Electrophor 1:73–76

Hochstrasser DH, Tissot JD (1993) Clinical applications of 2-D PAGE. In: Chrambach A, Dunn MJ, Radola BJ (eds) Advances in Electrophoresis. VCH, Germany (vol 6) pp 270–375

Hsich G, Kenney K, Gibbs CJ, Lee KH, Harrington MG (1996) The 14-3-3 brain protein in cerebrospinal fluid as a marker for transmissible spongiform encephalopathies. N Engl J Med 335:924–930

Hughes GJ, Frutiger S, Paquet N, Ravier F, Pasquali C, Sanchez JC, James R, Tissot JD, Bjellqvist B, Hochstrasser DF (1992) Plasma protein map: an update by microsequencing. Electrophoresis 13:707–714

Isoda N, Kajii E, Ikemoto S, Kimura K (1990) Two-dimensional polyacrylamide gel electrophoretic pattern of duodenal tumour proteins. J Chromatogr 534:47–55

James RW, Hochstrasser D, Tissot JD, Funk M, Appel R, Barja F, Pellegrini C, Muller AF, Pometta D (1988) Protein heterogeneity of lipoprotein particles containing apolipoprotein A-I without apolipoprotein A-II and apolipoprotein A-I with apolipoprotein A-II isolated from human plasma. J Lipid Res 29:1557–1571

Ji H, Moritz RL, Reid GE, Ritter G, Catimel B, Nice E, Heath JK, White SJ, Welt S, Old LJ, Burgess AW, Simpson RJ (1997a) Electrophoretic analysis of the novel antigen for the gastrointestinal-specific monoclonal antibody, A33. Electrophoresis 18:614–621

Ji H, Reid GE, Moritz RL, Eddes JS, Burgess AW, Simpson RJ (1997b) A two-dimensional gel database of human colon carcinoma proteins. Electrophoresis 18:605–613

Keim D, Hailat N, Hodge D, Hanash SM (1990) Proliferating cell nuclear antigen expression in childhood acute leukemia. Blood 76:985–990

Kovarova H, Stulik J, Hochstrasser DF, Bures J, Melichar B, Jandik P (1994) Two-dimensional electrophoretic study of normal colon mucosa and colorectal cancer. Appl Theor Electrophor 4:103–106

Kyte J, Doolittle RF (1982) A simple method for displaying the hydropathic character of a protein. J Mol Biol 157:105–132

Lawson SR, Latter G, Miller DS, Goldstein D, Naps M, Burbeck S, Teng NN, Zuckerkandl E (1991) Quantitative protein changes in metastatic versus primary epithelial ovarian carcinoma. Gynecol Oncol 41:22–27

Leroux JC, Gravel P, Balant L, Volet B, Anner BM, Allemann E, Doelker E, Gurny R (1994) Internalization of poly(D,L-lactic acid) nanoparticles by isolated human leukocytes and analysis of plasma proteins adsorbed onto the particles. J Biomed Mater Res 28:471–481

Lin JD, Huang CC, Weng HF, Chen SC, Jeng LB (1995) Comparison of membrane proteins from benign and malignant human thyroid tissues by two-dimensional polyacrylamide gel electrophoresis. J Chromatogr B Biomed Appl 667:153–160

Linnekin D, Evans G, Michiel D, Farrar WL (1992) Characterization of a 97-kDa phosphotyrosyl protein regulated by multiple cytokines. J Biol Chem 267:23993–23998

Marzewski DJ, Towfighi J, Harrington MG, Merril CR, Brown P (1988) Creutzfeldt-Jakob disease following pituitary-derived human growth hormone therapy: a new American case. Neurology 38:1131–1133

Marzocchi B, Magi B, Bini L, Cellesi C, Rossolini A, Massidda O, Pallini V (1995) Two-dimensional gel electrophoresis and immunoblotting of human serum albumin modified by reaction with penicillins. Electrophoresis 16:851–853

Moule SK, Peak M, Thompson S, Turner GA (1987) Studies of the sialylation and microheterogeneity of human serum alpha 1-acid glycoprotein in health and disease. Clin Chim Acta 166:177–185

Myers TG, Anderson NL, Waltham M, Li G, Buolamwini JK, Scudiero DA, Paull KD, Sausville EA, Weinstein JN (1997) A protein expression database for the molecular pharmacology of cancer. Electrophoresis 18:647–653

Myers TG, Dietz EC, Anderson NL, Khairallah EA, Cohen SD, Nelson SD (1995) A comparative study of mouse liver proteins arylated by reactive metabolites of acetaminophen and its nonhepatotoxic regioisomer, 3'-hydroxyacetanilide. Chem Res Toxicol 8:403–413

Nasmyth K (1993) Control of the yeast cell cycle by the Cdc28 protein kinase. Curr Opin Cell Biol 5:166–179

Nathan C, Sporn M (1991) Cytokines in context. J Cell Biol 113:981–986

Okuzawa K, Franzen B, Lindholm J, Linder S, Hirano T, Bergman T, Ebihara Y, Kato H, Auer G (1994) Characterization of gene expression in clinical lung cancer materials by two-dimensional polyacrylamide gel electrophoresis. Electrophoresis 15:382–390

Osada T, Sakamoto M, Ino Y, Iwamatsu A, Matsuno Y, Muto T, Hirohashi S (1996) E-cadherin is involved in the intrahepatic metastasis of hepatocellular carcinoma. Hepatology 24:1460–1467

Pun T, Hochstrasser DF, Appel RD, Funk M, Villars-Augsburger V, Pellegrini C (1988) Computerized classification of two-dimensional gel electrophoretograms by correspondence analysis and ascendant hierarchical clustering. Appl Theor Electrophor 1:3–9

Rasmussen HH, Orntoft TF, Wolf H, Celis JE (1996) Towards a comprehensive database of proteins from the urine of patients with bladder cancer. J Urol 155:2113–2119

Rasmussen RK, Ji H, Eddes JS, Moritz RL, Reid GE, Simpson RJ, Dorow DS (1997) Two-dimensional electrophoretic analysis of human breast carcinoma proteins: mapping of proteins that bind to the SH3 domain of mixed lineage kinase MLK2. Electrophoresis 18:588–598

Reymond MA, Sanchez JC, Schneider C, Rohwer P, Tortola S, Hohenberger W, Kirchner T, Hochstrasser DF, Kockerling F (1997) Specific sample preparation in colorectal cancer. Electrophoresis 18:622–624

Richardson FC, Horn DM, Anderson NL (1994) Dose-responses in rat hepatic protein modification and expression following exposure to the rat hepatocarcinogen methapyrilene. Carcinogenesis 15:325–329

Rush RS, Derby PL, Smith DM, Merry C, Rogers G, Rohde MF, Katta V (1995) Microheterogeneity of erythropoietin carbohydrate structure. Anal Chem 67:1442–1452

Sanchez JC, Wirth P, Jaccoud S, Appel RD, Sarto C, Wilkins MR, Hochstrasser DF (1997) Simultaneous analysis of cyclin and oncogene expression using multiple monoclonal antibody immunoblots. Electrophoresis 18:638–641

Sarto C, Marocchi A, Sanchez JC, Giannone D, Frutiger S, Golaz O, Wilkins MR, Doro G, Cappellano F, Hughes G, Hochstrasser DF, Mocarelli P (1997) Renal cell carcinoma and normal kidney protein expression. Electrophoresis 18:599–604

Saunders FK, Sharrard RM, Winfield DA, Lawry J, Goepel JR, Hancock BW, Goyns MH (1993) 2D-gel analysis of proteins in chronic lymphocytic leukemia cells and normal B-lymphocytes. Leuk Res 17:223–230

Schmid HR, Schmitter D, Blum P, Miller M, Vonderschmitt D (1995) Lung tumor cells: a multivariate approach to cell classification using two-dimensional protein pattern. Electrophoresis 16:1961–1968

Schwalke MA, Doremus CM, Bleday R, Wanebo HJ, Vezeridis MP (1990) Protein differences in human pancreatic cancer cell lines with diverse metastatic potential. Arch Surg 125:469–471

Spector NL, Mehlen P, Ryan C, Hardy L, Samson W, Levine H, Nadler LM, Fabre N, Arrigo AP (1994) Regulation of the 28 kDa heat shock protein by retinoic acid during differentiation of human leukemic HL-60 cells. FEBS Lett 337:184–188

Strohman RC (1993) Ancient genomes. Wise bodies. Unhealthy people: limits of genetic thinking in biology and medicine. Perspect Biol Med 37:112–145

Strohman RC (1994) Epigenesis: the missing beat in biotechnology? Bio/Technology 12:156–164

Strohman RC (1995) Linear genetics, non-linear epigenetics: complementary approaches to understanding complex diseases. Integr Physiol Behav Sci 30:273–282

Strohman RC (1997) The coming Kuhnian revolution in biology. Nat Biotechnol 15:194–200

Stulik J, Bures J, Jandik P, Langr F, Kovarova H, Macela A (1997) The different expression of proteins recognized by monoclonal anti-heat shock protein 70 (hsp70) antibody in human colonic diseases. Electrophoresis 18:625–628

Takahashi N, Breitman TR (1991) Retinoylation of proteins in leukemia, embryonal carcinoma, and normal kidney cell lines: differences associated with differential responses to retinoic acid. Arch Biochem Biophys 285:105–110

Takahashi N, Breitman TR (1992a) The covalent labeling of proteins by 17 beta-estradiol, retinoic acid, and progesterone in the human breast cancer cell lines MCF-7 and MCF-7/AdrR. J Steroid Biochem Mol Biol 43:489–497

Takahashi N, Breitman TR (1992b) Covalent modification of proteins by ligands of steroid hormone receptors. Proc Natl Acad Sci USA 89:10807–10811

Takahashi N, Breitman TR (1994) Retinoylation of vimentin in the human myeloid leukemia cell line HL60. J Biol Chem 269:5913–5917

Takahashi N, Liapi C, Anderson WB, Breitman TR (1991) Retinoylation of the cAMP-binding regulatory subunits of type I and type II cAMP-dependent protein kinases in HL60 cells. Arch Biochem Biophys 290:293–302

Tissot JD, Hohlfeld P, Forestier F, Tolsa JF, Hochstrasser DF, Calame A, Plouvier E, Bossart H, Schneider P (1993) Plasma/serum protein patterns in human fetuses and infants: a study by high-resolution two-dimensional polyacrylamide gel electrophoresis. Appl Theor Electrophor 3:183–190

Tissot JD, Hochstrasser DF, Schneider B, Morgenthaler JJ, Schneider P (1994) No evidence for protein modifications in fresh frozen plasma after photochemical treatment: an analysis by high-resolution two-dimensional electrophoresis. Br J Haematol 86:143–146

Ward LD, Hong J, Whitehead RH, Simpson RJ (1990) Development of a database of amino acid sequences for human colon carcinoma proteins separated by two-dimensional polyacrylamide gel electrophoresis. Electrophoresis 11:883–891

Watters DJ, Michael J, Hemphill JE, Hamilton SE, Lavin MF, Pettit GR (1992) Bistratene A: a novel compound causing changes in protein phosphorylation patterns in human leukemia cells. J Cell Biochem 49:417–424

Weinstein JN, Myers TG, O'Connor PM, Friend SH, Fornace AJ, Jr., Kohn KW, Fojo T, Bates SE, Rubinstein LV, Anderson NL, Buolamwini JK, van Osdol WW, Monks AP, Scudiero DA, Sausville EA, Zaharevitz DW, Bunow B, Viswanadhan VN, Johnson GS, Wittes RE, Paull KD (1997) An information-intensive approach to the molecular pharmacology of cancer. Science 275:343–349

Wieser RJ, Baumann CE, Oesch F (1995) Cell-contact mediated modulation of the sialylation of contactinhibin. Glycoconj J 12:672–679

Wilkins MR, Hochstrasser DF, Sanchez JC, Bairoch A, Appel RD (1996a) Integrating two-dimensional gel databases using the Melanie II software. Trends Biochem Sci 21:496–497

Wilkins MR, Sanchez JC, Williams KL, Hochstrasser DF (1996b) Current challenges and future applications for protein maps and post-translational vector maps in proteome projects. Electrophoresis 17:830–838

Wimmer K, Kuick R, Thoraval D, Hanash SM (1996) Two-dimensional separations of the genome and proteome of neuroblastoma cells. Electrophoresis 17:1741–1751

Wirth PJ, Egilsson V, Gudnason V, Ingvarsson S, Thorgeirsson SS (1987) Specific polypeptide differences in normal versus malignant human breast tissues by two-dimensional electrophoresis. Breast Cancer Res Treat 10:177–189

Witzmann FA, Fultz CD, Wyman JF (1997) Two-dimensional electrophoresis of precision-cut testis slices: toxicologic application. Electrophoresis 18:642–646

Yamawaki T, Toyoda N (1994) Human chorionic gonadotropin secretion and protein phosphorylation in chorionic tissue. Endocr J 41:509–516

Young DC, Wiehle RD, Joshi SG, Poindexter AN 3rd (1994) Emergency contraception alters progesterone-associated endometrial protein in serum and uterine luminal fluid. Obstet Gynecol 84:266–271

Young DS, Tracy RP (1995) Clinical applications of two-dimensional electrophoresis. J Chromatogr A 698:163–179

Yuasa I, Ohno K, Hashimoto K, Iijima K, Yamashita K, Takeshita K (1995) Carbohydrate-deficient glycoprotein syndrome: electrophoretic study of multiple serum glycoproteins. Brain Dev 17:13–19

Yun M, Wu W, Hood L, Harrington M (1992) Human cerebrospinal fluid protein database: edition 1992. Electrophoresis 13:1002–1013

Zahnd A, Funk M, Vaudaux P, Lew D, Scherrer JR, Hochstrasser DF (1994) A classification of one-dimensional electrophoresis gels using wave packet decomposition. Appl Theor Electrophor 4:19–24

Zahnd A, Tissot JD, Hochstrasser DF (1993) Wave packets analysis of two-dimensional protein maps: a new approach to study the diversity of immunoglobulins. Appl Theor Electrophor 3:321–328

Zerr I, Bodemer M, Otto M, Poser S, Windl O, Kretzschmar HA, Gefeller O, Weber T (1996) Diagnosis of Creutzfeldt-Jakob disease by two-dimensional gel electrophoresis of cerebrospinal fluid. Lancet 348:846–849

9 Biological Applications of Proteomics

Keith L.Williams and Vitaliano Pallini

9.1 Scope of this chapter

Although proteomics indicates a new view of mass protein analysis, it is in fact a synthesis of much that has been developing over the past twenty five years. In Chap. 2 through 7 of this book the status of the technology has been discussed, and how the different elements are integrated in proteome studies has been enunciated in Chap. 1. In Chap. 8 the emerging impact of proteomics in medicine has been considered. Here our brief is to take a broader, more biological view of where proteomics is (and will be) impacting on biology in general, particularly in areas of biotechnology that transcend medicine. It is important to realise that the field of proteomics is opening up as this book is being written. Thus much of the existing impact of proteomics involves use of the technology that is limited to 2-D gel mapping of complex protein mixtures. It is not surprising that to date there has been little characterisation of the proteins displayed, as only now are dedicated proteome facilities being established. However, current results do give a glimpse of the dramatic impact that proteomics is likely to have on all fields of biological sciences in the future.

9.2 Proteome maps

Just as a sequenced genome is the starting point for a major study in genomics, a completely defined proteome will be the knowledge base from which proteomics studies will expand. Some proteome studies will be relatively simple, for example the proteomes of viruses, mycoplasmas and other simple organisms. Even proteomes of prokaryotes are likely to be relatively uncomplicated, although there will probably be significant changes in the expressed genes and thus the proteome will alter with changes in growth conditions. However, for eukaryotes, the story will be quite different. A complete proteome of a eukaryote will involve many reference 2-D maps, even for a single-celled eukaryote. Here we consider what is involved with establishing proteomes of simple organisms, prokaryotes and eukaryotes, and we discuss those organisms likely to be "first in the queue". This chapter then concludes with the kinds of proteome study that will drive the development of the proteome maps, and early benefits are outlined.

9.2.1 Mollicutes

With the completion of the genomes of a number of mycoplasmas and other simple organisms, there is an incentive to see how much of the genome is transcribed and translated in the living organism, and indeed what effect different growth conditions have on the proteome. In the case of *Mycoplasma genitalium* the translated genome of 470 proteins has been theoretically imaged on a 2-D gel map to indicate where protein spots are predicted to lie (assuming no modifications occur to the proteins). To date no complete proteome 2-D maps are available, but a significant start has been made on *M. genitalium* (Wasinger et al. 1995) and *Spiroplasma melliferum* (Cordwell et al. 1995).

9.2.2 Prokaryotes

There has been a major effort to understand and map the proteome of *Escherichia coli* using 2-D gel display of cells labelled with [^{35}S]-methionine, with carrier ampholytes as the vehicle for the isoelectric focusing step (VanBogelen et al. 1992, 1996a). Recently this map has begun to be correlated with a 2-D map of *E. coli* using semi-preparative loading of gels prepared with IPG technology. Various techniques have been used to identify proteins in this study and some 200 have been mapped to specific spots on the 2-D gels (Pasquali et al. 1996). These early maps represent the foundation for important biological studies on how environmental stresses and starvation affect the proteome (VanBogelen et al. 1996b), and also for studies on the role of genes versus the environment on the expressed proteins (Nystrom and Neidhardt 1996).

Using IPG 2-D gel technology, a major effort is now under way to completely characterise the *E. coli* proteome by identifying proteins using a combination of N-terminal tag sequencing, amino acid analysis and peptide mass fingerprinting. In a pilot study some 150 proteins have been characterised (unpublished collaboration between the Hochstrasser (Geneva) and Williams (Sydney) groups). It is of interest that at least 15% of these *E. coli* proteins have been modified by N-terminal processing of the original protein, thus making clear that even prokaryote proteins are often modified. Similar results have been obtained in *Chlamydia trachomatis*, where the Major Outer Membrane Protein has a cleaved N-terminal signal, while Outer Membrane Protein 2 has both Mr and pI heterogeneity (Bini et al. 1996). Approximately 5% of *Chlamydia* proteins appear to be modified, based on the appearance of "trains" of spots on 2-D gels. Other less systematic studies make clear that considerable post-translational modification, including protein glycosylation (Tuomanen 1996), also occurs on prokaryotic proteins.

Clearly the initial work described above is setting the scene for application of proteomics to practical issues. *E. coli* is likely to be an important reference organism especially for gram negative bacteria as the genome of *E. coli* is now fully sequenced. Bacteria likely to be studied in the near future include the gram negative *Rhizobium* sp (for plant-bacteria interactions in relation to nodulation and nitrogen fixation) and *Pseudomonas* sp (pathogenic and non-pathogenic strains involved in, for example, problems with wearing contact lenses). Use of cross-species matching of proteins from these bacteria to *E. coli* proteins may be a powerful way to help identify proteins in these organisms whose genomes are not yet sequenced (see Sect. 3.5). The fully sequenced genome of *Bacillus subtilis* is likely to be a similarly useful reference for studies on gram positive bacteria.

On the other hand, crossing genus and species boundaries has its hazards. Recent studies on different serovars of the same species of *Chlamydia trachomatis* suggest that two serovars share only approximately 300 of 600 spots detectable on a 2-D map; the rest of the spots are sufficiently different not to be easily cross-referenced. In comparative studies between *C. trachomatis* and two other species of *Chlamydia* (*C. pneumoniae* and *C. psittaci*) only approximately 10% of spots could be confidently assigned as common landmarks between the species (Bini and coworkers, in preparation).

9.2.3 Eukaryotes

Since eukaryotes extensively modify their proteins by N- or C-terminal cleavages, and decorate them with sugars and/or phosphate, sulfate and a whole series of other modifications (see Chap. 4), any attempt to conduct a comprehensive proteome study on an eukaryote is likely to lead to the analysis of many more proteins than there are genes in the genome. Because the baker's yeast *Saccharomyces cerevisiae* is the first eukaryote whose genome has been fully sequenced, it is not surprising that many groups are focusing their initial major proteome studies on this organism. The ease of gene disruption and the hope that findings from yeast will translate directly to humans have also encouraged the emergence of several yeast proteome projects. Clearly some yeast genes will be common to those from humans; an initial estimate is that one in five known human disease genes have yeast counterparts (Richards 1997).

A European project is preparing 2-D maps of wild type and yeast strains with systematic gene deletions (Danish Proteome facility). A finding of the initial work is that deletion of a single gene leads to much more than just the loss of the protein which it encodes. Invariably other proteins disappear and others actually increase in amount. Still others may be differently modified. So the outcome of the loss of a single gene is a shift in the overall proteome of the organism. Organisms are buffered to counteract specific problems in their internal "wiring", without the result being lethal. However, there is a limit to this as some 20% of yeast genes encode essential proteins, the loss of which is lethal. Probably the most important outcome of the gene deletion studies will be to begin to understand the cross-talk and interactions between proteins in the yeast cell. Hence the networks of changes occurring will eventually be able to be put together in the form of trees of relationships between expressed proteins and protein linkage maps (Evangelista et al. 1996). Some of these affinities will reflect physical relationships between proteins (i.e. those proteins that form multiprotein complexes), but others will reflect information flow pathways, whereby the existence of a given protein influences the production or absence of production of other protein(s). Hence we predict an important outcome of proteome studies in the future will be to guide studies on protein-protein interactions as we build up a view of how cells are constructed and how their proteins work together.

In the USA several groups use yeast as their model for developing proteome analysis. Jim Garrels (Proteome Inc.) has constructed the Yeast Proteome Database (see Sect. 5.2.4) that encompasses much of what is known about this organism. Another group at the University of Washington is using yeast to develop their technology to undertake large scale proteome analysis, including studies on protein post-translational modifications (Figeys and Aebersold 1997).

Yeast is a simple eukaryote, with only a few different life stages. Simplistically there are three cell types: haploids of mating types "alpha" and "a", and sexual diploids. However, if the life of a yeast cell is to be looked at closely there are numerous different stages in growth, and also division of the cell by "budding" of a daughter cell. So it is possible to imagine perhaps 15 proteomes (i.e. 3 cell types, each with say 5 different stages), although the reality is that there will be a continuum between different stages. Finally, the physiological state of a cell is controlled by environmental factors, so the medium in which the cell is grown, ambient temperature and availability of oxygen will have an influence on the proteome. The changes in the proteome seen in different growth conditions are likely to be variations on a general theme, but how this complexity can be displayed in a manner that is comprehensible and meaningful to the researcher remains a formidable bioinformatics problem.

Some would argue that studies on *S. cerevisiae* will keep scientists busy for many years to come, but scientists are all different, and some will wish to tackle even more complex, less well defined challenges. Perhaps the next logical step after studies on yeast is to choose an organism on the boundary between single celled and multicellular organisms. The cellular slime mould *Dictyostelium discoideum* is such an organism. It has many of the advantages of yeast, including small genome size (probably 8,000–12,000 genes) and genetic manipulation so that, as in yeast, any gene can be disrupted, and it can be grown under well-defined conditions. *D. discoideum* is a single-celled amoeba, not unlike a macrophage, in its growth phase. When starved, the amoebae signal each other and form a multicellular tissue by aggregation of cells. Hence initially solitary amoebae form a mound of perhaps 100,000 cells. Subsequently, in a tightly controlled sequence of developmental events, three basic cells types are differentiated and a tiny, motile slug-like organism is formed. This undergoes a period of migration before finally constructing a fruiting body consisting of spores, a cellular stalk and cellular basal disc. Perhaps 10 developmental stages can be identified and proteomes of each stage can be prepared (Raper 1984). The organism can also be viewed from a cellular perspective, and proteomes could be constructed on the three basic cell types as they develop. Of course the world is not so simple and it is likely that perhaps as many as ten cell types are formed and these will require study as the means to identify and purify them develops. The organism also makes an extracellular matrix which surrounds the multicellular stage and through which the cells move in the migratory phase (Wilkins and Williams 1995). *D. discoideum* has a number of the standard co- and post-translational modifications found on animal proteins, and so this organism can be used as a pilot for studies on, for example, glycoforms. All of the different developmental stages can be subjected to proteome analysis and a start has recently been made on this project with the preparation of a 2-D map of the "slug" stage (Yan et al. 1997).

It can be anticipated that all of the model organisms whose genomes are being sequenced will be subjected to proteome analysis in the near future. Already several different ways of approaching proteome analysis in organisms with complex life cycles are becoming apparent. Bini et al. (1997) have used an "averaging" approach to studying the proteome of the nematode worm *Caenorhabditis elegans*. They took a mixed culture containing a range of developmental stages and displayed this "average" nematode on a 2-D map, finding over 2,000 spots in the apparent size range 10–200 kDa and isoelectric focusing window of pI 3.5–9. Such a study gives a generalised picture that sets the foundation for studies on individual life stages.

On the other hand Ericsson et al. (1997) studied the adult (fly) stage of the life cycle of *Drosophila melanogaster*. Flies of both sexes were dissected into head, thorax and abdomen and 2-D maps were prepared for each body part. Most of the 1200 proteins detected were common to all three body parts, but each part had a small group of body-part and/or sex-specific proteins.

In a third approach Rouquie et al. (1997) constructed a 2-D map of plasma membrane proteins purified from tobacco *Nicotiana tabacum* suspension cultures. Here the focus of the study was on the proteins that interact with the environment and hence are responsible for initiation of signalling responses. Some 600 or 80% of the expected proteins were displayed.

In all of the above studies most of the proteins encountered remain to be identified, and the application of methods described in the earlier chapters of this book are clearly needed to uncover the vast amount of information that is currently hidden.

There is too much interest and investment in human biology to wait for the results of pilot proteome studies, so considerable effort is already going into human proteome analysis (see Chap. 8). Because of the large number of tissues, cell types and developmental stages, studies on humans are highly focused on specific tissues, cell types or diseases. This work suggests that many proteins (perhaps 80%) found in some human tissues (such as liver and kidney) are shared "house-keeping" proteins, despite the fact that these tissues have quite different functions (Dean et al. 1994). This contrasts to the different developmental stages of some simpler organisms that have radically different proteomes. For example, studies on different life stages of the truffle fungus *Tuber borchii* show markedly different 2-D maps (Vallorani et al. 1996). Thus current proteome studies are already beginning to reveal interesting differences in how simple and complex organisms achieve specialisation, hence defining how we can best study the proteins of these organism types. In humans, one well-characterised 2-D map will be useful as a reference for understanding many other tissues. However in a fungus, we may need to study many different life stages before a good definition of the organism's proteome is reached.

9.3 Applications of proteome technology

9.3.1 Tracking complexity

Many studies in biology involve trying to untangle complex issues. Examples include host-pathogen or host-parasite interactions. These can be beneficial, as in the case of nitrogen fixation in legumes by association with bacteria (*Rhizobium* sp) to form nodules. There are a number of ways that such studies can be approached. Different bacterial isolates have different capacities to form nodules and fix nitrogen, so proteome analysis of such strains may give insight into molecules that are important for nitrogen fixation. Alternatively, one might study plant roots with and without nodulation to understand both plant-related and bacteria-related proteins involved with nodule formation. In a pilot study, Geurreiro et al. (1997) have identified a number of the products of *nod* genes in their model system. *Nod* genes in *Rhizobium* are those known to be involved with nodulation. Finding such proteins gives confidence in proteome analysis technology, and the discovery of new proteins will guide the molecular biologists in their gene cloning studies.

Other researchers at the Australian National University are studying the process of host-pathogen interactions using the model of infection of flax by the flax rust. There is clear evidence of proteins differentially expressed in pathogenic isolates of the fungus and proteome studies are under way to find the nature of these proteins. Whether the proteins prove to be key fungal proteins or plant defence molecules, the important point is that through a proteome study a way into understanding a complex biological issue has been found.

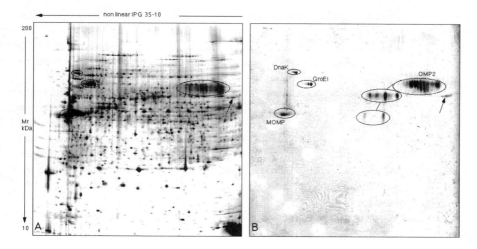

Fig. 9.1. Exploiting prokaryote proteome information to dissect human antibody responses to infection. (A) 2-D electrophoretic display of *Chlamydia trachomatis* proteins visualised by silver-staining within the experimental window of Mr 10–200 kDa and pI 3.5–10. Some spots have been assigned as reported in Bini et al. (1996). (B) Western-blotting of a similar 2-D gel blot with serum from a woman with infection by *C. trachomatis*. Targets of her antibody response are detected by peroxidase-labelled secondary antibody. These spots were matched to the silver-stained pattern in A using Melanie II software, as reported in Bini et al. (1996). In this clinical case, immunoreactive spots include isoelectric series of heat shock proteins DnaK and GroEl, Major Outer Membrane Protein (MOMP), Outer Membrane Protein 2 (OMP2) (circled in figure) and an unidentified component (arrow). The complex post-translational pattern of OMP2, which includes proteolytic processing, is similarly evidenced by artificially raised monoclonal antibodies (Bini et al. 1996)

9.3.2 Immunogenic proteins

There are a number of applications where display of the products of an entire genome or tissue will allow a global appreciation of a biological phenomenon. An excellent example is in identifying proteins from infectious disease agents recognised by the immune system. When a person contracts an infectious disease the immune system responds to various parts of the infecting organism. By displaying the proteome of the infectious organism and using serum from people infected by this organism, one can identify the major immunogenic proteins (Fig. 9.1). Clearly those proteins which are highly immunogenic are potential vaccine candidates for microbial pathogens. We expect that this may become a standard approach to studying infectious disease. Initial studies on *Chlamydia trachomatis* infection in humans have revealed candidate antigens which have been identified using the proteome approach (Ratti et al. manuscript in preparation). Immuno-reactivity is an important feature of a protein and such information needs to be collected and made easily available through annotation on 2-D maps. We expect an explosion of such valuable information to be a key benefit from the use of 2-D gel mapping experiments.

Another area where knowledge about the immunogenicity of proteins is crucial is in the field of allergy research. Knowing which grass pollens are most immunogenic may lead to the development of desensitising treatments for people with severe grass allergies (Petersen et al. 1997), and in the case of crop plants, the long-term development of plants with low allergenicity (Weiss et al. 1997). Proteome research has already been applied to an unusual allergy case. Latex gloves used widely in medicine can be allergenic to some users, posing problems where gloves are needed as a contact barrier. Posch et al. (1996) studied the proteins in latex gloves to see if they could understand which ones were the cause of allergy. Although latex gloves are not an obvious starting material for 2-D gel electrophoresis, the researchers were able to extract proteins and identify the key allergens in a relatively straightforward manner.

9.3.3 Improved agricultural products

Traditionally agriculture has been an empirical discipline, although in the area of plant and animal breeding there was a highly structured approach even before the rules of genetics were understood. In the past decade agriculture has entered the biotechnology revolution with the introduction of transgenic technology. This is exemplified by engineering resistance to pathogens/parasites (viral, bacterial, fungal or insect) into various plants. Most of these resistance mechanisms involve the production of toxic or protective proteins, and expression of the protein, as opposed to high levels of mRNA, is essential for protection (e.g. resistance in potatoes (Tacke et al. 1996)). It will be very surprising if proteomics is not used soon to aid discovery of new toxic proteins (such as those from *Bacillus thuringiensis* of which there are at least a hundred (Estruch et al. 1997)) or to identify plants likely to be resistant to a pathogen, that is, expressing high levels of the relevant protein, at an early stage after transformation.

There are many examples of agricultural products where the nature of valuable traits is not understood at the molecular level and the phenotype is likely to result from several proteins rather than a single gene product. In the case of wool, aspects such as colour and fibre strength are economically important characters. With the definition of a wool proteome (Herbert et al. 1997), it should be possible to identify proteins related to these characters and hence to develop rational breeding programs or molecular biological approaches to improve the product.

9.3.4 Value added agricultural products

With the internationalisation of agricultural markets and pressure on efficiency, there is a need to review all aspects of processing of agricultural products. By-products formerly seen to be of low value are now being re-examined to see if there is potential for new product development, and old products are being reviewed to see if it is possible to add value to them by processing. Three areas under pressure are dairy products, abattoir wastage and cereals. In each of these areas of agriculture, there are very large tonnages of relatively low-value products. In the past, significant by-products of the processing were either disposed of or sold to cover disposal costs. It may seem surprising that detailed analysis of proteins in such low technology biotechnology areas has utility, but a brief overview of these three technologies indicates that this is indeed the case.

In the dairy industry, proteinaceous whey is a by-product of cheese manufacture. Current research programs using proteome technology are seeking to find growth factors from such material in the anticipation that these proteins will be useful in various areas of biotechnology. They could for example have application as growth promoters for recombinant protein production, or for use in the development of wound healing agents. Treatment of ulcers in the elderly is a major health issue and a product which had utility in this area would have considerable value and may change the economics of cheese making!

Abattoirs traditionally collect the blood, bone and offal from slaughter of animals and produce a plant fertiliser called "blood and bone". Serum contains at least 2,000 different proteins of which only a few hundred are known. A major proteome study of serum is likely to lead to the discovery of many useful proteins that could be used to enhance the value of this by-product of the meat industry.

The food industry is increasingly focused on processed foods and such products require a source of bland proteins. Traditionally caseins from milk have been a major source of such protein, but casein is relatively expensive. A source of appropriate protein from a cereal would have a significant market. But before such products can be developed it is important to understand the nature of the starting material so that a rational means of product development can be instigated. Two dimensional gel technology is an obvious way of establishing such a knowledge base on the raw protein extracts obtained from cereals.

9.3.5 Quality control

We are living in times where regulation is a feature of almost all human activities. With regulation come issues of product definition and specification. By specifying products governments can accept or reject imports, lawyers can establish the basis for defence of liability or prosecute on the basis of failure of an aspect of the product. It is an issue whether the hamburger mince sold as beef really is beef or a mixture of beef and kangaroo or even buffalo. Better specifications mean more clarity for both the producer and consumer. Proteome technology brings precision and definition to a new level in protein-based products. Not only is each protein fully specified (including whether or not each protein is truncated or has post-translational modifications) but the composition of a product containing many proteins can become apparent from one or more 2-D gels. Rumour has it that the first legal dispute involving 2-D gel identification of a plant variety has recently been heard.

Many industries would argue that protein technology has been too complex and skill-intensive to be practical for use in quality specification in the past. This book makes clear that this is no longer the case, and readily accessible 2-D display technology is now available to most laboratories. It is true that complete proteome analysis is still a challenging and capital-intensive process, but most quality control processes require a much simpler definition of the product. This can currently be achieved in a straightforward manner.

We anticipate that proteome specification will soon be required for both high value (e.g. recombinant proteins for therapeutic use or vaccines) and low value (e.g. protein products used as food additives) products. Importantly, this will include details about the status of modifications to the protein(s), especially glycosylation status. Such specification will not be limited to the products, but will also include the components for manufacturing products. For example the cell line for making recombinant products is likely to be specified with reference to a published map (e.g. a certain chinese hamster ovary cell line as defined in the SWISS-2DPAGE database). This provides a benchmark not only for the genetic features of the cell line, but also for the particular growth conditions, since it is clear that culture conditions affect the nature of post-translational modifications in mammalian expression systems (Goochee and Monica 1990).

Other highly specific features of biotechnology products will be quality controlled by proteome technology. A likely example is the development of 2-D gel-based methods for detecting the presence of bovine spongiform encephalopathy (BSE or "mad cow") disease in cattle. Since there is evidence that "mad cow" disease is communicable to humans, where it is manifested as a new strain of CJD (Creutzfeldt-Jakob Disease), there is much pressure to develop an assay in living cattle for existence of the prion protein causing the disease. A recent report suggests that the new prion variant has a different glycosylation pattern in comparison with other known forms of BSE or CJD prion proteins (Hoyle 1997).

9.4 Concluding remarks

This chapter has scratched the surface of applications for proteome technology beyond the confines of the pharmaceutical and biomedical industries. We can be sure of but one thing, that our view into the crystal ball is limited in the extreme. We expect all areas of biology to feel the impact of proteome technology in the near future.

References

Bini L, Sanchez-Campillo M, Santucci A, Magi B, Marzocchi B, Comanducci M, Christiansen G, Birkelund S, Cevenini R, Vretou E, Ratti G, Pallini V (1996) Mapping of *Chlamydia trachomatis* proteins by Immobiline-polyacrylamide two-dimensional electrophoresis: spot identification by *N*-terminal sequencing and immunoblotting. Electrophoresis 17:185–190

Bini L, Heid H, Liberatori S, Geier G, Pallini V, Zwilling R (1997) Two-dimensional gel electrophoresis of *Caenorhabditis elegans* homogenates and identification of protein spots by microsequencing. Electrophoresis 18:557–562

Cordwell SJ, Wilkins MR, Cerpa-Poljak A, Gooley AA, Duncan M, Williams KL, Humphery-Smith I (1995) Cross-species identification of proteins separated by two-dimensional gel electrophoresis using matrix-assisted laser desorption time of flight mass spectrometry and amino acid composition. Electrophoresis 16:438–443

Dean DP, Cronan MT, Merenda JM, Gardner JP, Connelly PA, Celis JE (1994) "Spot transfer", elution and comigration with known proteins allows accurate transferral of protein identifications between distinct two-dimensional electrophoretic systems. Electrophoresis 15:540–543

Ericsson C, Pethö Z, Mehlin H (1997) An on-line two-dimensional polyacrylamide gel electrophoresis protein database of adult *Drosophila melanogaster*. Electrophoresis 18:484–490

Estruch JJ, Carozzi NB, Desai N, Duck NB, Warren GW, Koziel MG (1997) Transgenic plants: an emerging approach to pest control. Nat Biotechnol 15:137–141

Evangelista C, Lockshon D, Fields S (1996) The yeast two-hybrid system: prospects for protein linkage maps. Trends Cell Biol 6:196–199

Figeys D, Aebersold R (1997) High sensitivity identification of proteins by electrospray ionization tandem mass spectrometry: initial comparison between an ion trap mass spectrometer and a triple quadrupole mass spectrometer. Electrophoresis 18:360–368

Geurreiro N, Redmond JW, Rolfe BG, Djordjevic MA (1997) New *Rhizobium leguminosarum* flavonoid-induced proteins revealed by proteome analysis of differentially displayed proteins. Mol Plant Microbe Interact 10:506–516

Goochee CF, Monica T (1990) Environmental effects on protein glycosylation. Bio/Technology 8:421–427

Herbert BR, Molloy MP, Yan JX, Gooley AA, Bryson WG, Williams KL (1997) Characterisation of wool intermediate filament proteins separated by micropreparative two-dimensional electrophoresis. Electrophoresis 18:568–572

Hoyle R (1997) The link between Creutzfeldt-Jakob disease and BSE. Nature Biotechnol 15:295

Nystrom T, Neidhardt FC (1996) Effects of overproducing the universal stress protein UspA, in *Escherichia coli* K-12. J Bacteriol 178:927–930

Pasquali C, Frutiger S, Wilkins MR, Hughes GJ, Appel RD, Bairoch A, Schaller D, Sanchez JC, Hochstrasser DF (1996) Two-dimensional gel electrophoresis of *Escherichia coli* homogenates: the *Escherichia coli* SWISS-2DPAGE database. Electrophoresis 17:547–555

Petersen A, Grobe K, Lindner B, Schlaak M, Becker WM (1997) Comparison of natural and recombinant isoforms of grass pollen allergens. Electrophoresis 18:819–825

Posch A, Chen Z, Wheeler C, Dunn MJ, Raulf-Heimsoth M, Baur X (1996) Characterization and identification of latex allergens by two-dimensional electrophoresis and protein microsequencing. In: From Genome to Proteome, Abstracts, 2nd Siena 2-D Electrophoresis Meeting, Siena, Italy, September 16–18, 1996

Raper KB (1984) The Dictyostelids. Princeton University Press, New Jersey USA

Richards W (1997) Saccharomyces sapiens. Trends Genet 13:49–50

Rouquie D, Peltier JB, Marquis-Mansion M, Tournaire C, Doumas P, Rossignol M (1997) Construction of a directory of tobacco plasma membrane proteins by combined two-dimensional gel electrophoresis and protein sequencing. Electrophoresis 18:654–660

Tacke E, Salamini F, Rohde W (1996) Genetic engineering of potato for broad-spectrum protection against virus infection. Nat Biotechnol 14:1597–1601

Tuomanen EI (1996) Surprise? Bacteria glycosylate proteins too. J Clin Invest 98:2659–2660

Vallorani L, Bernardini F, Zambonelli A, De Belis R, Ceccaroli P, Saltarelli R, Bini L, Pallini V, Stocchi V (1996) 2-D PAGE pattern of *Tuber borchii* Vitt. mycelium, *Tilia Platyphyllos* roots and ectomycorrhiza. In: From Genome to Proteome, Abstracts, the 2nd Siena 2-D Electrophoresis Meeting, Siena, Italy, September 16–18, 1996

VanBogelen RA, Sanker P, Clark RL, Bogan JA, Neidhardt FC (1992) The gene-protein database of *Escherichia coli*: edition 5. Electrophoresis 13:1014–1054

VanBogelen RA, Abshire KZ, Pertsemlidis A, Clark RL, Neidhardt FC (1996a) Gene-Protein Database of *Escherichia coli* K-12: Edition 6. In: Neidhardt FC, Curtiss RI, Gross CA, Ingraham JL, Riley M (eds) *Escherichia coli* and *Salmonella typhimurium* Cellular and Molecular Biology, 2nd Edition. ASM Press, Washington, pp 2067–2117

VanBogelen RA, Olson ER, Wanner BL, Neidhardt FC (1996b) Global analysis of proteins synthesized during phosphorus restriction in *Escherichia coli*. J Bacteriol 178:4344–4366

Wasinger V, Cordwell SJ, Cerpa-Poljak A, Gooley AA, Wilkins MR, Duncan M, Williams KL, Humphery-Smith I (1995) Progress with gene-product mapping of the mollicutes: *Mycoplasma genitalium*. Electrophoresis 16:1090–1094

Weiss W, Huber G, Engel KH, Pethrun A, Dunn MJ, Gooley AA, Görg A (1997) Identification and characterization of wheat grain albumin/globulin allergens. Electrophoresis 18:826–833.

Wilkins MR, Williams KL (1995) The extracellular matrix of the *Dictyostelium discoideum* slug. Experientia 51:1189–1196

Yan JX, Tonella L, Sanchez JC, Wilkins MR, Packer NH, Gooley AA, Hochstrasser DF, Williams KL (1997) The *Dictyostelium discoideum* proteome — the SWISS-2DPAGE database of the multicellular aggregate (slug). Electrophoresis 18:491–497

10 Conclusions

Denis F. Hochstrasser and Keith L.Williams

10.1 Scope of this chapter

So far we have conveyed a sense of a new and vibrant field of biology, and pointed to many of the new paths that will indicate highways of the future. In this chapter we briefly revisit where proteomics fits in the overall scheme of biology, and also consider possible future scenarios and some alternative paths down which new avenues of research might go.

10.2 Large-scale science: genomics, combinatorial chemistry and proteomics

The past two decades have seen revolutionary developments in many areas of biology and chemistry. The discovery of the means for manipulating and amplifying DNA has made study of DNA and RNA facile. Genes can be sifted from the masses of genomic sequence. They can be cloned and inserted into expression vectors. The proteins that they encode can now be routinely engineered in a bacterial or eukaryote host be it microbe, plant or animal (or in cell lines derived from such a host). Miniaturisation of pumps, valves and switches in conjunction with the computer revolution has made possible the development of microanalytical instruments of exquisite sensitivity and throughput. New chemistries coupled with various detectors (including the mass spectrometer) have played their part in this instrument development. Computers, fast microchips and networks have allowed the development of the field of bioinformatics, imaging, artificial intelligence and a means of world-wide communication, the World-Wide Web.

The above fundamental discoveries have encouraged massive investment in new biotechnology ventures. In particular the field of genomics and associated major projects, such as the human genome project, have blossomed. This started as a series of massive DNA sequencing initiatives, but has now broadened to include studies on genotyping, gene expression, gene disruption and functional studies on gene disruptants. The commercial goal of such business ventures is in areas of drug discovery and finding new proteins related to specific diseases or other complex biological processes (Friedrich 1996). Looking at this from the other end of the problem (i.e. structure biology), the field of combinatorial chemistry has, like genomics, attracted massive interest and investment by the pharmaceutical companies (Hogan 1997).

It is interesting that genomics and combinatorial chemistry each indirectly address proteins. Genomics comes from an informational perspective, while combinatorial chemistry is aimed at producing compounds that are agonists or antagonists for specific proteins. Proteomics is an obvious link between these two major areas of biotechnology, as it addresses the functional molecules (proteins) directly. Proteomics also opens up the issues of protein modification that cannot currently be described from a genomics perspective, and which to date are not well understood from a structure biology perspective.

By combining genomics, proteomics, and combinatorial chemistry with robotics and parallel sample handling, a very powerful suite of tools becomes available not only to the biotechnologist for drug discovery, but also to all aspects of biology. These are exciting times.

10.3 Future methodologies in proteomics

Now that proteome is in the psyche, it is very hard to imagine that the new view of large-scale protein analysis could be lost. Hence there seems little possibility of a return to the view of protein science as a slow and difficult art.

In this book we are totally committed to multidimensional and parallel protein separations and we have focused particularly on the two-dimensional (2-D) gel as the most practicable way of achieving separation of large numbers of proteins (Chap. 2). Others are less convinced and would use microchips to display the proteins, or a series of liquid chromatography or capillary electrophoresis steps coupled to mass spectrometry and/or MS/MS. Each of these technologies has considerable promise. But there are still significant technical hurdles which must be overcome for them to be competitive with the 2-D gel as a general screening method. For example, to our knowledge the reproducible separation and resolution of proteins of more than 30 kDa in a complex mixture is yet to be achieved by capillary electrophoresis. In addition, despite the exquisite detection sensitivity of laser induced fluorescence, there is a problem with detecting rare proteins by capillary electrophoresis because of the extremely small sample volumes which are separated by this type of technology. The difference between the most and least concentrated protein in a cell is in the order of magnitude of 6 logs (i.e. perhaps between 100 and 100 million molecules), although in body fluids this difference will be larger. Therefore techniques are required which can separate abundant proteins and concentrate rare ones. Narrow-range 2-D gels are well suited to this process.

There are mathematical and natural laws which describe the world as we see it. In mathematics, when a differential equation is too complex, the approach is to use a combination of a general solution and a particular case solution. The full analysis of a very complex biological sample such as a proteome may require a general methodology and a different approach for specific questions. In nature, two physico-chemical parameters, mass and charge, are key features of matter. Interestingly, the periodic table of elements is organised by charge, (one to eight outer electrons), and by nucleus and atomic size (one to more than 100). The concept of separating proteins by charge and size follows natural laws. In addition, like in mathematics, a focusing methodology is an attractive way of achieving reproducible separation.

Orthogonal and distinctly unrelated separation techniques must provide excellent resolution power. Ideally, each dimension should involve a focusing step. Immobilised pH gradients are today the most reproducible methodology with the highest loading capacity to separate polypeptides by charge. It is a focusing technique with fixed engineered pH gradients. Pore limit separations in PAGE are promising for providing an array of separated proteins in the second dimension. Alternatively, mass spectrometry may provide distinction of different masses as a second dimension.

Theoretically, a 3-D technique could be designed where the first dimension would be a hydrophobicity separation using a chromatography method, then a charge separation in an IPG plate by electrofocusing and finally scanning the IPG plate with an IR-MALDI-TOF MS for the accurate mass determination. The 3-D image would have a hydropathy value for the x coordinate, a charge value for the y axis and a mass determination for the z dimension. Three-dimensional imaging software would match and cluster these 3-D images. All dimensions would be orthogonal with minimal inter-correlation. The resolving power of the first dimension (hydrophobicity separation) should be about one third of IEF or SDS-PAGE (therefore 30). The IEF separation has a resolution of more than 100. Scanning IR-MALDI-TOF MS has a resolution of more than 100. This 3-D technique would thus have a resolution of $30 \times 100 \times 100 = 300,000$! We would most likely have the resolution required to detect all gene products, including post-translationally modified forms.

Getting back to earth, with any methodology, a bottleneck is the loading capacity and the separation of poorly soluble proteins in the first dimension. With the development of new methods for solubilising insoluble proteins coupled with an array of narrow-range first dimension strips for the 2-D gel separations, the answer may be already at hand (Chap. 2). There will, however, always be a role for tissue subfractionation. Why try to find plasma membrane proteins amongst the cytoplasmic proteins, when cell fractionation using centrifugation provides more than a 100 fold enrichment of the proteins of interest?

10.4 Other views of the future of proteomics

As indicated in Chapter 3, an attribute-based approach to protein identification is inevitably the way to approach protein identification and characterisation. We already believe that the utility of technologies that are directed solely at protein identification is limited. Hence amino acid analysis or peptide mass fingerprinting on their own are of limited value. Already the focus is on protein characterisation, which includes (but is not yet generally feasible) N- and C-terminal identification, as well as getting an accurate mass for the whole protein. The major problem for the success of a mass screening approach to protein characterisation is to avoid the necessity for moving from parallel processing at the separation (2-D gel) stage, to serial processing of individual protein spots. When this bottleneck is solved, proteomics will truly have become a technology that will be complementary to genomics, not only in its power of discovery, but also in its throughput capacity. We await the discovery of such technology with interest.

Chapter 4 addresses areas where other technologies (e.g. genomics) have little to say. The strategy outlined is focused on multiple analysis of modifications being conducted sequentially on the same protein spot separated by 2-D gel technology. The practicability of such an approach was demonstrated, but it must be acknowledged that there are still some problems with sensitivity in doing such work. However, only one further logarithmic increase in sensitivity is required.

Chapters 5, 6 and 7 address fascinating bioinformatics questions. Clearly the ways of classifying data and displaying both DNA and protein information are now well in hand. The central remaining questions are storage, annotation and quality controlled access to the exponentially growing planetary information repository, and linking sequence information to protein structure and function. The ultimate dream is to be able to identify a protein from any species, to automatically do inter-species comparison and postulate its structure and function.

The value of SWISS-PROT (Chap. 5) is the quality of its annotation and the minimum amount of redundancy as well as the cross-references to numerous databases world-wide. The ExPASy server has been built around SWISS-PROT as its heart (Chap. 6). The annotation of SWISS-PROT provides the required information to dramatically improve the protein characterisation tools of ExPASy, such as the MultiIdent program. SWISS-PROT also provides the data which is the basis for protein modelling programs described in Chap. 7. The progress in protein modelling has been astonishing. The combination of databases and tools described in Chap. 5 to 7 provides one of the most powerful and integrated virtual biological laboratories available on the Internet today.

Chapters 8 and 9 address the here and now of new territory being explored in proteomics. We believe that proteome research will lead to the development of clinical molecular scanners which will unravel new diagnostic markers or patterns. Proteome research will lead to the development of better prognostic evaluation and new therapeutic modalities. Proteins are considered in the same way as genes for current proteome studies. Hence, because of the use of denaturing conditions, multisubunit proteins are seen as a series of individual peptide chains. It requires only a little imagination to foreshadow the emergence of a parallel native 2-D gel system for studying protein complexes.

Proteomics today is largely about making biology better defined, and providing the next generation of researchers with a catalogue of some of the ingredients of life. We must, nevertheless, remember that this is just the beginning of understanding how organisms function.

In biology, we are now at the stage of assembling the pieces of the puzzle. This could be thought of as learning about the pieces in the game of chess. It is exciting to learn that there are 64 squares, 16 pawns, 4 rooks and 2 queens, but this is in no way comparable to knowing how to play. Life still has many secrets. We hope readers, especially young people starting a career in science, have captured our excitement and fun as new frontiers open before us. And we hope that some may be stimulated to join in the interesting times!

References

Friedrich GA (1996) Moving beyond the genome projects. Nat Biotechnol 14:1234–1237
Hogan JC (1997) Combinatorial chemistry in drug discovery. Nat Biotechnol 15:328–330

Index

Springer
and the
environment

At Springer we firmly believe that an
international science publisher has a
special obligation to the environment,
and our corporate policies consistently
reflect this conviction.
We also expect our business partners –
paper mills, printers, packaging
manufacturers, etc. – to commit
themselves to using materials and
production processes that do not harm
the environment. The paper in this
book is made from low- or no-chlorine
pulp and is acid free, in conformance
with international standards for paper
permanency.

Springer